A WORL

Cali, Colombia
Toward a City Development Strategy

The World Bank
Washington, D.C.

© 2002 The International Bank for Reconstruction and Development / The World Bank
1818 H Street, NW
Washington, DC 20433

All rights reserved.
1 2 3 4 04 03 02

World Bank Country Studies are among the many reports originally prepared for internal use as part of the continuing analysis by the Bank of the economic and related conditions of its developing member countries and of its dialogues with the governments. Some of the reports are published in this series with the least possible delay for the use of governments and the academic, business and financial, and development communities. The typescript of this paper therefore has not been prepared in accordance with the procedures appropriate to formal printed texts, and the World Bank accepts no responsibility for errors. Some sources cited in this paper may be informal documents that are not readily available.

The findings, interpretations, and conclusions expressed here do not necessarily reflect the views of the Board of Executive Directors of the World Bank or the governments they represent.

The World Bank cannot guarantee the accuracy of the data included in this work. The boundaries, colors, denominations, and other information shown on any map in this work do not imply on the part of the World Bank any judgment of the legal status of any territory or the endorsement or acceptance of such boundaries.

Rights and Permissions

The material in this work is copyrighted. No part of this work may be reproduced or transmitted in any form or by any means, electronic or mechanical, including photocopying, recording, or inclusion in any information storage and retrieval system, without the prior written permission of the World Bank. The World Bank encourages dissemination of its work and will normally grant permission promptly.

For permission to photocopy or reprint, please send a request with complete information to the Copyright Clearance Center, Inc., 222 Rosewood Drive, Danvers, MA 01923, USA, telephone 978-750-8400, fax 978-750-4470, www.copyright.com.

All other queries on rights and licenses, including subsidiary rights, should be addressed to the Office of the Publisher, World Bank, 1818 H Street NW, Washington, DC 20433, USA, fax 202-522-2422, e-mail pubrights@worldbank.org.

ISBN: 0-8213-5174-5
ISSN: 0253-2123

Library of Congress Cataloging-in-Publication Data has been applied for.

Cover photo by: Alejandro Ardila, El Puente Ortiz, símbolo de la ciudad de Cali.

CALI, COLOMBIA: TOWARD A CITY DEVELOPMENT STRATEGY

CONTENTS

ABSTRACT vi
PREFACE vii
ACKNOWLEDGMENTS ix
CURRENCY EQUIVALENTS, FISCAL YEAR, ACRONYMS AND ABBREVIATIONS x

INTRODUCTION
Objectives of the City Development Strategy 1
Counterpart team 1
Methodology 1
Data sources 3
The report 3

1. CALI IN CONTEXT
Colombia's socioeconomic crisis 5
Colombian urban development 5
Cali within the Colombian context 6
Cali's territorial organization 7
Cali's administrative organization 8
Cali's municipal government strategy 8
Other city programs 10

2. INSTITUTIONAL MODERNIZATION
Introduction 12
Governance in Cali before 1990 12
The challenges of governance in present-day Cali 16
Proposed strategic priorities for institutional modernization 22

3. ECONOMIC REACTIVATION
Introduction 24
Overview of Cali's economy 24
Cali's labor market 26
Economic sectors 28
Exports 34
Infrastructure 37
Proposed strategic priorities for Cali's economic recovery 39

4. SOCIAL DEVELOPMENT
Introduction 42
Cali in the national urban perspective, 1994-1998 42
Poverty in Cali 44
Municipal programs I: expenditure distribution and access 52
Municipal programs II: satisfaction with basic and human service provision 54
Proposed strategic priorities for social development in Cali 56
Annex A. Regression Results for the Wage Regression 59

5. URBAN VIOLENCE
Introduction 60
The national context 60

    Crime and violence in Cali 61
    Relationship between violence and social, institutional, political, and legal aspects 66
    Current programs for violence reduction in Cali 69
    Strategic priorities for violence reduction in Cali 72

6. EDUCATION
    Introduction 76
    Education in Colombia 76
    Education in Cali 78
    Proposed strategic priorities for education in Cali 83

7. FINANCIAL SITUATION
    Introduction 86
    The central administration 86
    The municipal companies 92
    The Metro project 94
    Proposed strategic priorities to solve the municipal finance crisis 95

8. SUMMARY OF FINDINGS
    Cali: a city in a severe crisis 98
    Proposals for the future 98

ANNEXES
    A. Map of Cali 103
    B. Cali at a glance 104
    C. Note on spatial management 105

BIBLIOGRAPHY 106

TABLES
    1.1. Homicide rates in Colombia, 1976-1999 5
    1.2. Population and density in Cali, 1992-1997 7
    2.1 Cali: city development and city governance, 1900-1990 13
    2.2 Cali: participation rates in community affairs, September 1999 19
    2.3 NGOs affiliated with PROCALI, by sector 21
    2.4 Projects jointly executed by NGOs and the municipal government, 1998 22
    3.1. Structure of employment in Cali, 1990/94 and 1995/98 27
    3.2. Occupational status of the labor force in Cali, 1990/94 and 1995/98 27
    3.3. Unemployment in Cali by sector, 1997 27
    3.4. Characteristics of the labor market in Cali, September 1999 29
    3.5. Cali's industrial structure, 1980-84 and 1995-97 30
    3.6. Employment informality and education level in Cali, 1998 34
    3.7. Employment informality by economic activity in Cali, 1998 35
    3.8. The changing structure of manufactured exports, Valle del Cauca Department, December 1992-1996 36
    4.1. Cali in the national perspective 43
    4.2. Unemployment in Cali, 1994 and 1998 43
    4.3. Social security access in major Colombian cities 43
    4.4. Poverty and misery in Cali, 1994 and 1998 44
    4.5. Income and socioeconomic strata, Cali, 1999 45
    4.6. Poverty headcount rates by *comuna* 46
    4.7. Rank order of *comunas* by poverty, extreme poverty, and number of poor, 1999 47
    4.8. Characteristics of income poverty, Cali, 1999 51
    4.9. Composition of municipal expenditures, 1997 52
    4.10 Incidence of education and health expenditures in Cali, September 1999 54
    4.11 Dissatisfaction with basic social and infrastructure services, Cali, September 1999 55
    4.12 Priorities for expansion of municipal programs, Cali, September 1999 (percent) 56

4.13 Priorities for cut-back of municipal programs, Cali, September 1999 (percent)............ 56
5.1. Reported crime and homicide rates in Colombia and the Cali metropolitan area, 1990-1998 62
5.2. Contribution to Cali's homicides by most violent neighborhoods, 1996 64
5.3. Summary of existing programs for violence reduction in Cali 70
6.1. School performance in Cali and Medellin, 1999 80
6.2. Education and poverty in Cali, 1999 81
6.3. Unemployment levels by educational attainment in Cali, March 1999 82
6.4. Yearly private rates of return to education in Cali, September 1999 82
7.1. Breakdown of expenditures by activity, 1998 87
7.2. Breakdown of number of employees, 1999 88
7.3. Breakdown of current revenues, 1998 88
7.4. Investment program, 1999 89
7.5. The Colombian regulatory framework: borrowing limits 90
7.6. Municipal companies, basic indicators 93

BOXES

1.1. The city development process 2
4.1. EPSOC: Encuesta de acceso y percepción de los servicios ofrecidos por el Municipio de Cali 45
5.1. Deterioration of networks of cooperation 66
5.2. Mistrust between the police and the community 67

GRAPHS

1.1. Cali's administrative organization 9
2.1. Rent-seeking model in Cali Twentieth century 17
2.2. Willingness of the population to participate in government programs 18
3.1. Real growth rate of GDP: Valle del Cauca vs Colombia 25
3.2. Main sectors of Cali's economy 26
3.3. Participation of industrial sectors in Cali's GDP, 1990-1996 31
3.4. Degree of employment informality in Cali, 1986-1998 34
4.1. Poverty and age, Cali 1999 50
4.2. Social expenditure distribution against poverty distribution, 1997 53
4.3. Transport expenditure distribution against poverty distribution, 1997 53
4.4. Other expenditure distribution against poverty distribution, 1997 53
5.1. Homicide rates in Cali metro area, 1980-1998 62
5.2. Relationship between homicides and poverty 65
5.3. Perception of police performance, 1999 68
6.1. School enrollment rates in Colombia, 1985-1997 77
6.2. Educational attainment in Colombia, 1995 78
6.3. Third grade language achievement scores in Latin America, 1999 79
6.4. School participation rates of children 7-11 and 12-17 in five Colombian cities, 1997 79
6.5. Attendance in Cali primary schools, children ages 6-12, 1994-1998 80
7.1. Breakdown of total expenditures by type, 1998 87
7.2. Breakdown of local taxes, 1991-1998 89
7.3. Financial Adjustment Plan, 1998-2003: current revenues and expenditures before interest 91
7.4. Financial Adjustment Plan, 1998-2003: financial expenses 91
7.5. Financial Adjustment Plan, 1998-2003: capital revenues and investments 92
7.6. Financial Adjustment Plan, 1998-2003: debt 92
7.7. Financial Adjustment Plan, 1998-2003: debt including and excluding the METRO project 94

MAPS

4.1. Poverty in Cali, 1999 headcount rates 48
4.2. Distribution of poverty, 1999 49
5.1 Homicide rates per *comuna* in Cali, 1997 63

ABSTRACT

Although many of the problems that Cali is experiencing, social and human capital deterioration, a declining economy, and an institutional crisis, are a reflection of Colombia's complicated socioeconomic situation, the city has been hit harder by the crisis than other large cities as confirmed by the following indicators: GDP, unemployment, poverty rate, inequality and number of homicides. According to recent estimates, the population in Cali reached the 2 million level in 1999, making the city the second largest in the country after Bogotá. It has been growing at an average 1.83 percent per year in the period 1994-1999. Internal migration increased significantly during the early 90s due to the economic boom generated by drug dealing activities, and continued in the last part of the decade, due to resettlement movements of large population groups affected by social conflict in rural areas. However, these later migration flows have generated social tension in the city, as economic opportunities became scarce in the latter part of the decade.

In order to help the city, the World Bank engaged in a participatory process to produce a City Development Strategy (CDS). Specifically, the objectives of the Cali CDS are: (i) to help the city administration and main stakeholders in identifying a strategy to overcome the present city crisis; and (ii) to be a neutral facilitator in a much needed city leadership reconstruction process. The CDS is being developed in four stages: (a) city-wide identification and prioritization of the main problems of the city; (b) development of the analytical framework; (c) dissemination and validation of results; and (d) financing plan. Stages (a) and (b) have been completed and have as output this report. Stage (c) has been developed to a certain extext, however it might be taken a step further along with the development of Stage (d) as a municipal responsibility.

PREFACE

Over the past decade, few of the fastest growing cities in Latin America have been able to capitalize on the promises of global trade. The majority struggle to catch up, to bring service levels in line with population demand. Ambitious reforms of decentralization and governance, undertaken in most Latin American countries, have not been enough to make cities livable, competitive, well governed, efficiently managed, and financially sustainable. Cities still face many difficulties, particularly with the allocation of resources, the timing and planning of investments, and the coherence of basic service delivery. The result is that many cities are unable to realize their full growth potential, large segments of urban populations are still living in poverty, and much of the region's infrastructure is in need of repair.

An urban transition is underway in the developing world, signifying an additional two billion urban residents in the next quarter century. Meeting the needs of renewal, growth, and reduction in urban poverty is increasingly both a national and local matter, as cities and nations are swept up in the tides of globalization, democratization, and decentralization. These forces expose cities as never before to outside competition, to the imperatives of democratic local governance, and to a much more important role in national affairs.

The City Development Strategy, like the one elaborated here for Cali, is a new tool in development assistance, a tool designed to respond to the tidal changes sweeping the globe, and to make urban poverty alleviation and local economic development the centerpieces of the development puzzle. More than 80 CDSs are now underway or have been completed in a dozen countries. CDSs aim to blend participatory inputs from all segments in society—as Cali did in an innovative way—in order to reach an integrated long-term vision for the city, establish a growth strategy, agree on priorities for solutions, and identify short-term actions. Most CDSs, like that of Cali, benefit from Cities Alliance financing.

Cali presented a stark set of circumstances at the time the city requested World Bank assistance to elaborate its CDS in 1998. The city was inextricably caught up in the nation's recession, its drug trade, and its civil war, and suffered from unemployment and inflation, both in double digits. The city was prostrate in economic and financial terms, having lost its economic base and hopelessly in debt. At the same time, local leadership was focused on a large-scale rapid transit system project as a way to integrate the city in a physical sense. In psychological and other ways, it was thought that the transit project might pull the fractured city together around a large common undertaking, even though the most rudimentary financial analysis showed that a metro rail system in Cali was utterly nonviable. In the first year of work on the CDS, many months of group interaction were devoted to synthesizing the findings of economic, social, financial, and physical dimensions of the city. When a new administration took office and fell into step with the process, it rejected the idea of a metro project and instead focused on laying the groundwork to refinance its debt. The many working groups representing neighborhoods, workers, NGOs, the business community, and others found ways to then begin to focus on other means to rebuild. The working groups focused issues of violence and security, the most dire problems of poverty, and on reversing the flight of businesses in order to create jobs. The story is still far from complete in Cali, but a sharp turn has been taken and a new direction established.

The process of elaborating the CDS in Cali has also contributed to national efforts. In acquiring a sense of coherence and control in its own development efforts, cities such as Cali are in a stronger position to respond to the demands of decentralized democracies. In this sense, the CDS represents something of a precedent for the country. The planning process in Cali has shown not only how cities can shoulder some of the burden of reform, but also how they can take on fiscal and service responsibilities and contribute to national objectives. Bogotá and other cities have begun to take steps in this same direction, and Bank management is exploring ways to scale up these efforts to a national level. Similar steps are also being taken in Mexico, China, and Vietnam, where CDSs are also being elaborated in major cities. In these and other countries, a national strategic view can harness individual city overtures, drawing on the experiences such as Cali's, draw on their cities' CDS experiences to create a more cohesive response to the many challenges for nations and cities in decentralized democracies and a globalized economy as they become more decentralized, democratic, and integrated into the global economy .

Tim Campbell
World Bank Urban Lead Specialist
December, 2001

ACKNOWLEDGMENTS

This report is the outcome of a collaborative process between the World Bank and the city of Cali.

The World Bank's task team included: Alexandra Ortiz (Task Manager, LCSFU), Eleoterio Codato (co-task manager, LCSFU), Laura Tagle (former LCSFU), Jairo Arboleda (LCC1C), Felipe Saez (LCCVE), Maria Mar (SRMIG), Kene Ezemenari and Jesko Hentschel (PRMPO), Fernando Rojas (LCSPS), Tim Campbell and Pilar Solans (INFUD), Edgar Bueno (Colombian National Planning Department), and Nigel Harris, Gerard Martin, and Benjamin Alvarez (consultants). Administrative assistance was provided by Jose Herdoiza (former LCSFU), Antonio Trivizo, and Jason McGee (LCSFU). Peer reviewers were: Christine Kessides (INFUD), Geoffrey Shepherd (LCSPS), and Haeduck Lee (LCSPP). Comments on a prior draft were received from peer reviewers, local counterparts in Cali, George Gattoni (INFUD), Mila Freire and Dean Cira (LCSFU), Clemente Luis del Valle (FSD), Carlos Eduardo Velez (LCSPP), Zeinab Partow (LCC1C), Juan Gaviria (LCSFP), and Elsie Garfield (LCSER). Many other colleagues contributed with advice and guidance, particularly, Danny Leipziger (LCSFP), Jeffrey S. Gutman (LCSFT), John Stein (LCSFU), and Mark Hildebrand (INFUD). Final editing was provided by Deborah Davis.

The report was compiled and edited by Alexandra Ortiz. Background papers were prepared by team members as follows:
- "Institutional Reform in Cali," by Fernando Rojas, with contributions from Alexandra Ortiz and Melba Pinedo
- "Toward an Economic Strategy for the City of Santiago de Cali," by Nigel Harris
- "Poverty in Cali, Colombia," by Jesko Hentschel, with contributions from Kalpana Mehra and Radha Seshagiri
- "Crime and Violence in Cali, Colombia—A Diagnosis and Policy Propositions," by Gerard Martin
- "City of Cali Development Strategy—The Education Perspective," by Benjamin Alvarez
- "The Municipality of Cali: Financial Review," by Pilar Solans, with contributions from Eleoterio Codato

The Cali team included: Ricardo Cobo (Mayor), Juan Carlos Narváez, Fabiola Perdomo, Myriam Hormaza, Francisco Mejía, Esneda Mogollón, Edgar Ivan Ortiz, Alvaro Tobón, Carlos Alberto Roldán, Fabiola Aguirre, Ana Maria Rojas, Florencia Lince, Jorge Medina, Luis Felipe Sánchez, Amparo Valencia, Ivette Martínez, Ivette García, Ruby Cuellar, Catalina Rebolledo, Fanny Herrea, Lilia Hincapié (Municipal government staff), Julián Dominguez, Maria Elena Suarez, Oscar Echeverri, Daniel Zamorano, Ricardo Bermúdez, Oscar Echeverri (Chamber of Commerce), Melba Pinedo (Association of NGOs, PROCALI), Fernando Urrea, Carlos Ortiz, Alvaro Guzmán, Pedro Martínez (Universidad del Valle), Fabio Velásquez (FORO por Colombia), Hernando Toro (Defensor del Pueblo), Gral. Jorge E. Montero (Head of the Cali Metropolitan Police), José González (Sacerdote de la Diócesis Arzobispal), Clara Luz Roldán (DESEPAZ), Maria Eugenia Pérez (Environment Department)

Funding for the study was provided by: The Colombia Country Unit (LCC1C), the Cities Alliance Program, the Strategic Compact, the Municipal Finance Thematic Group, the Dutch Trust Fund, and the United Kingdom's Department for International Development (DFID-UK).

CURRENCY EQUIVALENTS
US1$= 2292.50 Colombian Pesos

FISCAL YEAR
January 1 to December 31

ACRONYMS AND ABBREVIATIONS

BANCALI:	Municipal Bank of Cali
CDS:	City Development Strategy
CALI	Center of Integrated Local Administration
CALIASFALTO:	Cali Municipal Asphalt Plant
CIAT:	International Center for Agriculture and Technology
CIDSE:	Socioeconomic Research Center at Universidad del Valle
CISALVA:	Center for Violence Research at Universidad del Valle
DANE:	Colombian Statistical Department
DAP:	Cali City Planning Department
DESEPAZ:	Municipal Program for Development, Security, and Peace
DNP:	National Planning Department
EMCALI:	Cali Municipal Services Company
EMSIRVA:	Cali Solid Waste Collection Company
EPSOC:	Household Survey to Measure Access and Satisfaction with the Municipal Services of Cali
FORO:	Colombian NGO (*Foro Nacional por Colombia*)
ICBF:	Colombian Family Institute
IDB:	Inter-American Development Bank
ISS:	Colombian Social Security Institute
JAC:	Community Level Boards (*Juntas de Acción Comunal*)
JAL:	Local Administrative Board (*Juntas Administradoras Locales*)
MSSP:	Cali Municipal Strategy for Security and Peace
PROCALI	Cali Association of NGOs
SFEC:	Cali Economic Development Secretariat
SISBEN:	Colombian System of Beneficiary Information for Targeted Social Programs

Vice President:	David de Ferranti
Country Director:	Olivier Lafourcade
Sector Director:	Danny Leipziger
Task manager:	Alexandra Ortiz
Co-task manager:	Eleoterio Codato

INTRODUCTION

Objectives of the City Development Strategy

The City Development Strategy (CDS) is a non-lending advisory service instrument aimed at a comprehensive analysis of key urban issues in a particular city, and at helping define strategic options and implementation alternatives. Since CDSs are at a pilot stage in the Bank, each case has unique approaches and methodologies. The objectives of the Cali CDS are to: (i) help the city administration and main stakeholders to identify a strategy to overcome the present city crisis; and (ii) help facilitate a much-needed process of rebuilding the city's leadership. The Cali CDS has also developed a series of tools for urban analysis that the Bank can use in urban sector work and urban project preparation.

It is important to stress that the Cali CDS is both a <u>participatory process</u> and an <u>analytical study</u>, as will be explained in the Methodology section below.

Counterpart team

The Cali CDS is being prepared through a highly collaborative process. The main counterparts are the Cali Municipal Government, the Chamber of Commerce, and PROCALI, the city's association of NGOs. Other agencies have participated in particular activities, among them the NGO *Foro Nacional por Colombia,* the Socioeconomic Research Center of the *Universidad del Valle* (CIDSE), the local university ICESI, and the Colombian National Planning Department (DNP).

Methodology

The CDS is being developed in four stages.

Stage I. City-wide identification and prioritization of the city's main problems
Following a highly participatory and innovative approach, six priority issues were identified by Caleños as the most pressing: (a) economic reactivation; (b) social development; (c) urban violence; (d) education; (e) institutional modernization; and (f) spatial management. These issues framed the work on the CDS. It might be argued that other issues, including transport, housing, and health care are also crucial for the city, but it would have been impossible to cover all such issues with this study's limited time horizon and budget.

Stage II. Development of the analytical framework
Teams of Bank and local experts were formed to study each issue in depth, using primary and secondary data. The teams also thoroughly discussed alternative solutions. The output of this stage is represented in chapters 2 through 7 of this report.

Stage III. Dissemination and validation of results
This stage, which is to be developed by Caleños, includes the Spanish translation of the Bank document, the production of a synthesized and easy-to-read summary, the dissemination of this summary among citizens, and its discussion in focus groups. The result of this stage should be a strategy for action that will include specific investments.

It is expected that the city will take the lead in Stage III, with the Bank playing an advisory role, particularly in terms of participatory processes and follow-up.

Stage IV Financing plan

Once a strategy is defined with the citizens of Cali, the Bank might work with local authorities and private sector groups to define a feasible financial strategy for implementing the chosen options.

Box 1.1. The city development process

City-wide identification and prioritization of problems

The first step in the Cali CDS was a participatory consultation process to identify and prioritize city problems. A total of 250 people from different sectors and groups in Cali were invited to participate in the consultation meetings that took place on April 7-9, 1999. Of those invited, 147 participated (see table). There was good representation by gender and age group as well.

This consultation was not a statistically representative survey, which would have been quite expensive and time consuming, but rather a way to gauge the level of agreement on which of the city's problems are most pressing. The consultation used 16 computer terminals, a central server, and Groupware software, which allows for: (i) simultaneous registration of multiple entries; (ii) anonymity; (iii) equal value for all answers; and (iv) immediate calculation of results.

Consultation participants

Group	Number of participants
NGOs	27
Citizens	25
Municipal government, central	23
Private sector	13
Municipal government, decentralized (incl. many community leaders)	12
Academia (universities)	6
Sectoral/professional associations	6
Departmental government	6
Youth groups	6
Academia (schools & institutes)	4
Others	21
Total	147

The consultation took place over six separate sessions. Each session comprised five steps: (i) each participant identified two main challenges for Cali in the years to come; (ii) the resulting challenges were organized by theme at a plenary discussion; (iii) for each theme, participants expressed its importance, how to approach it, and their views of the role of government, the private sector, and civil society; (iv) each participant ranked the themes according to their importance; and (v) each participant gave a final recommendation. The results were as follows:

Themes	Votes
Economic reactivation/employment	1,456
Social development/poverty alleviation, attention to vulnerable groups	983
Peaceful coexistence/safety/urban violence	935
Education	917
Institutional modernization/governance/corruption	818
Urban planning/spatial planning	489

There was an outstanding degree of consensus in the selection and ranking of issues, with economic reactivation and employment generation the most important issues in five of the six sessions. These results were similar to those obtained from recent consultations carried out by the municipality's Social Policy Committee.

Development of the analytical framework
To proceed to the next stage, groups comprising both local counterparts and Bank experts were formed for each of the six issues. The groups were intended to be inter-institutional, but in practice this was not always the case. In fact, groups had different dynamics and ways of operating, and some were more successful than others. Each group had working sessions during a week in Cali to review reports, interview key informants, and hold internal discussions. Each group made a presentation on September 17, 1999. The Bank experts, with input from their local counterparts, wrote the chapters containing the diagnostics (chapters 2 through 7) and initial proposals for solutions.

Source: Author's compilation, 1999.

Data sources

The diagnostics in this report benefited from multiple sources of information, listed in the bibliography. Specific studies were commissioned where necessary. These were:

- Quick review of national household survey results for the largest Colombian cities, carried out by the national government 1994-1998 by consultant Mauricio Santamaria.
- Desk review of poverty-related studies by the Socioeconomic Research Center at Universidad del Valle (CIDSE).
- Institutional map by the Socioeconomic Research Center at Universidad del Valle (CIDSE).
- Household survey to measure satisfaction with municipal services (Encuesta de Acceso y Percepción de los Servicios Ofrecidos por el Municipio de Cali, EPSOC) by the firm Centro Nacional de Consultoría. The survey covered 1,912 households in the city. The results of the survey are used throughout the report, and are explained in detail in chapter 4.
- Household survey to assess resettlement needs of a small community by the youth group Ashanty.
- Economic prospects for Cali, by the firm Misión Siglo XXI.

The report

The aim of the report is to summarize the analytical work carried out to date as part of the CDS process, and to present for further discussion a preliminary set of recommendations to help the city in its efforts to recover from the present crisis. For several reasons, this is not a conventional Bank report: (a) since it is the result of only part of the process (the consultation and diagnostic stages, Stages I and II), the recommendations are only a first sketch for a city strategy; (b) the amount and quality of secondary information varied a great deal for each topic, resulting in some chapters having a wealth of quantitative data to support the analysis—urban violence is one of these cases—while other chapters have little information or information of dubious quality; and (c) as a result, the chapters are uneven in terms of depth of analysis and maturity of recommendations. The primary research for the social development chapter was supported by external funding, making it possible to carry out household surveys and create an institutional map. Although there are cross-references among the chapters, each chapter can be read separately.

It is important to note that the recommendations laid out in each chapter are the result of the work of the thematic groups—that is, they are joint proposals of the World Bank and local counterparts, not solely of the World Bank. Although some local team members have now moved on to different positions, this document remains faithful to the work of the thematic groups that took place in Sepember 1999. It is hoped that once the document is officially delivered to our counterparts, it will form the basis for a dynamic working relationship, including with new members of the teams, during subsequent stages of the CDS process.

Even though Caleños did not select municipal finance as one of the most important themes for urban development, it is, in fact, a fundamental part of any city devlepment strategy, since it has a direct bearing on the feasibility of proposals. Therefore, a chapter on municipal finance has been included in this report. With regard to spatial management, that thematic group was dissolved prior to the Bank's September 1999 mission because of disagreements about how spatial management affects a number of other issues, including housing, transporation, and the environment. Thus, the discussion of spatial management appears as an annex rather than a chapter.

The report is organized as follows: Chapter 1 presents basic information about Cali and locates the city's problems in a national context. Chapters 2 through 7 present, for each issue, a detailed diagnostic and initial proposals for strategic options; and Chapter 8 summarizes the main findings and recommendations, making an effort to integrate the different issues and present a consolidated and comprehensive set of proposals. Annex A is a city map, Annex B contains a one-page summary of statistics on Cali for easy reference, and Annex C is a note on spatial management in Cali.

Although the institutional modernization issue ranked fifth in the city-wide consultation, it is presented first in the report because it provides important context for the rest of the document. The other chapters are presented according to the rankings given by participants.

The report addresses two distinct audiences: (a) the CDS team, including our local counterparts, and (b) the Bank, particularly staff working on Colombia and on urban projects in different regions.

1. CALI IN CONTEXT

Colombia's socioeconomic crisis

Colombia is a middle-income country with rich natural and human resources and a strong democratic tradition. Despite a long and complex armed conflict, the country has managed to maintain a solid and constant economic growth of about 4.5 percent a year over the last four decades. Economic stability, coupled with a drop in the population growth rate, enabled the country to achieve outstanding progress during that period, including a steady drop in the infant mortality rate, a decline in the incidence of poverty, an increase in the primary school net enrollment rate, and innovative reforms in education, health, labor, and municipal management.

In spite of its richness and steady progress, however, Colombia is today one of the most conflict-torn countries in Latin America. Until the 1990s, the conflict had affected mainly combatants and populations in remote rural areas and in marginalized city neighborhoods. During the last decade, however, the scale and intensity of violence reached astonishing levels, and became the worst problem for all Colombians. Table 1.1. shows the escalation of homicide rates, which peaked in the early 1990s, when the main drug cartels were prosecuted and dismantled.

Table 1.1. Homicide rates in Colombia, 1976-1999

Year	1976	1980	1985	1990	1995	1999
Homicide rates (number of homicides/100,000persons)	26	34	43	75	72	59

Source. Martin, 2000.

Also in the mid to late 1990s, Colombia's GDP began to decrease, due to: (a) large imbalances in the current and fiscal accounts; (b) adverse external shocks, including a drop in international commodity prices (coffee, petroleum), increased costs of borrowing abroad, and drying-up of capital inflows; and (c) insufficient competitiveness of local industries in the face of opening of the local economy and global competition. These conditions were amplified by the intensification of the armed conflict and generalized violence. As a result, the GDP fell by 4.5 percent in 1999 and unemployment reached 20 percent.

Colombian urban development

Colombia is a highly urbanized country, with approximately 74 percent of its population now living in urban areas. Rural-urban migration is continuing because of social conflicts in the countryside. The primacy of its capital city is not as strong as in most South American countries. In fact Colombia has developed a balanced system of cities, with Bogotá, Medellín, Cali, and Barranquilla all above the one million population mark.

The political, fiscal, and administrative decentralization that began in the 1980s is now in a process of consolidation. Decentralization has provided local governments, and notably large cities, with public elections of mayors and councilors for a period of three years, as well as organizational independence and full responsibility for their own financial management. Decentralization has had positive and negative effects. Positve effects include increased local participation and accountability, and strengthening of the regulatory and legal frameworks for municipalities. But decentralization has also resulted in a number of serious problems, as illustrated by the large number of bankrupt municipalities, and in some cases, decreased efficiency of service delivery. The problems have led to the presentation of two important structural reform bills in Congress. The first, just enacted, aims to control expenditures and rationalize the finances of local governments; and the second, a constitutional amendment still under consideration, would alter the formula used to calculate territorial transfers, which have been a significant drain on central government finances.

The decentralization process has stressed the importance of local governments preparing and implementing local development plans. To this end, the Law of Urban Reform (1989) provided municipal governments with urban land management mechanisms such as expropriation, designation of priority areas for urban expansion, land banks, land readjustment, land value capture, and transfer of construction and development rights. Subsequent laws strengthened planning processes at the city and reginal levels. For example, Law 99 (1993) established the mechanisms and guidelines for municipalities to prepare environmental plans; Law 128 (1994) set the conditions for the formation of metropolitan areas; Law 134 (1994) established mechanisms to strengthen citizen participation in municipal decisionmaking processes; and Law 388 (1997) amended the Law of Urban Reform by incorporating elements of recent environmental, metropolitan, and territorial legislation. It also established the requirement that all municipalities must design and implement spatial plans (*plans de ordenamiento territorial,* POT) that deal exclusively with land use, land regulation, land development, and city expansion. Spatial plans are intended to complement the local development plans. But unlike local development plans, which are short term and tied with local government election cycles, the spatial plans have a time frame of nine years and therefore cover three administrations. With all these legal actions, the Government of Colombia has been moving the urban agenda toward integrated urban planning as opposed to sector-specific planning, and toward continuity of management processes and high levels of citizen participation in urban decisionmaking.

Cali within the Colombian context

Cali is located in the southwest of Colombia, and is a center for economic growth in this geographic area. It is the capital city of the Valle del Cauca Department, a fertile and dynamic region; and is close to the country's main Pacific port, Buenaventura, which accounts for a great proportion of shipping exports and imports. Cali is also close to the borders of Ecuador and Peru, giving it a strategic role in commercial activities with these countries.

According to recent estimates, Cali's population reached the 2 million mark in 1999, making it the second largest city in the country, after Bogotá. It grew an average of 1.83 percent a year during 1994-1999.[1] Internal migration increased significantly during the early 1990s due to the economic boom generated by drug dealing activities, and continued in the last part of the decade due to resettlement movements of large population groups affected by social conflict in rural areas (*desplazados,* the displaced). In the period 1994-1999, migration represented 45 percent of the annual net population growth in Cali, and this percentage is expected to continue increasing. These migration flows have generated social tensions in the city, as economic opportunities, already scarce, have gotten thinner. The overall density in urban Cali was estimated at 156 inhabitants per hectare in 1997 (see table 1.2).

---

[1] Urrea, 1997.

Although many of Cali's problems—high unemployment, increasing homicide rates, and institutional crisis—are a reflection of Colombia's complicated socioeconomic situation, the city has been hit harder by the crisis than have other large cities. First, in economic terms, Cali's GDP, which had been growing at higher annual rates than the nation in the first half of the 1990s, decreased by 1.79 percent in 1995 and 4.69 percent in 1996, while Colombia's GDP grew by 5.96 and 2.07 percent in those years.[2] There is no recent information on Cali's GDP, but looking at other indicators, its performance seems to be continuing below the national average. Official unemployment in Cali was an estimated 20 percent in 1998, while the average unemployment figure for urban Colombia (six largest cities) was 16 percent. Similarly, Cali's poverty rate in 1998 was 2.5 percent above the average rate for urban Colombia.[3] Finally, in terms of violence, the homicide rate in Cali in 1999 was 95 per 100,000—better than the figure for Medellin (166 per 100,000) but worse than for Bogotá (39 per 100,000) and for Colombia as a whole (59 per 100,000).

Cali's territorial organization

Cali is organized into 20 urban divisions, called *comunas*, and 15 rural divisions, called *corregimientos*. A 21st *comuna* is being defined at the extreme west of the city. Urban growth is taking place mostly in the south and west. The north area of Cali is bound by its proximity to Yumbo, a highly industrialized town, while the east is constrained by mountains. Although there is neither a metropolitan authority nor a formal definition of a metropolitan area, it is commonly accepted that the metropolitan region is formed by six municipalities: Cali, Candelaria, Jamundí, Palmira, Yumbo, and Puerto Tejada (please refer to Annex A).

Table 1.2. Population and density in Cali, 1992-1997

Year	1992	1993	1994	1995	1996	1997
Municipality						
Population	1,759,139	1,801,820	1,843,506	1,886,360	1,927,025	1,964,068
Gross density (inhab/ha)	31.19	31.95	32.69	33.45	34.17	34.82
Comunas						
Population	1,714,363	1,748,815	1,776,438	1,812,876	1,842,222	1,866,256
Gross density	143.60	146.48	148.80	151.85	154.31	156.32
Corregimientos						
Population	25,943	26,573	27,226	27,903	28,567	29,215
Gross density	0.58	0.60	0.61	0.63	0.64	0.66
Population in expansion areas	18,833	26,432	39,842	45,581	56,236	68,597

Source: Departamento Administrativo de Planeación Municipal, Alcaldía de Santiago de Cali, 1999.

---

[2] Mision Siglo XXI, 1999.
[3] Santamaria, 1999.

Cali's administrative organization

The central administration in the municipality of Cali is organized in two management units: (a) sectoral development, which includes physical, social, and collective sectors; and (b) spatial development, a decentralized unit that includes 20 Centers of Integrated Local Administration (CALIs), with one CALI per communa. Three departments cut across these two management units: municipal planning, cadastre and treasury, and environment (please refer to Graph 1.1). The central administration had 9,174 employees in 1997, who represented annual salary expenses of approximately $94 million. The services provided by the central administration are urban planning and housing, health and education, social assistance, transport, thoroughfare, public safety, economic promotion, environment, culture, and sports.

The decentralized administrative organization is intended to promote a participatory and decentralized approach to city planning. City-wide policies are operationalized in each CALI according to its particular characteristics. At the same time, the city development plan is based on the local development plans of each CALI.

Cali's municipal government strategy

The strategy of the municipal government is articulated in its municipal development plan, which proposes 20 separate programs, each with a set of goals to be achieved and supporting activities. Highlights of the most relevant activities in each program are summarized below, but this account is by no means exhaustive. It is important to note that the light railway project, also known as the metro project, was the most important project of the 1998-2000 administration.

- *Economic development.* This program promotes the creation of export activities; improves road access from Cali to the port of Buenaventura and from Cali to the industrial suburb of Yumbo; provides and complements incentives to new companies to locate in the new Ley Paez industrial park; increases productiveness of existing industries; promotes foreign investment; and creates a comprehensive program to support research and training in science and technology.

- *Public finance.* This program improves information systems; simplifies procedures for the payment of taxes and services; restructures the debt with the national government; develops new sources of revenue; develops and implements policies for investment of municipal resources; rationalizes expenditures; improves risk management and tax cost analysis; and titles vacant land.

- *Institutional modernization.* This program improves working conditions and training of municipal government staff, sets up a system of indicators to make public management more transparent, and improves physical infrastructure.

- *Education and health.* This program improves physical infrastructure and the provision of materials in public schools. It also promotes programs, in association with NGOs, to expand coverage of primary and secondary education, revise curricula to conform with national education policies, promote the continuing education of teachers, consolidate citizen education programs, increase coverage of public health programs, strengthen the social security institute, develop control mechanisms for private health care providers, improve health information systems, and increase community participation in public health management.

Graph 1.1 Cali's administrative organization

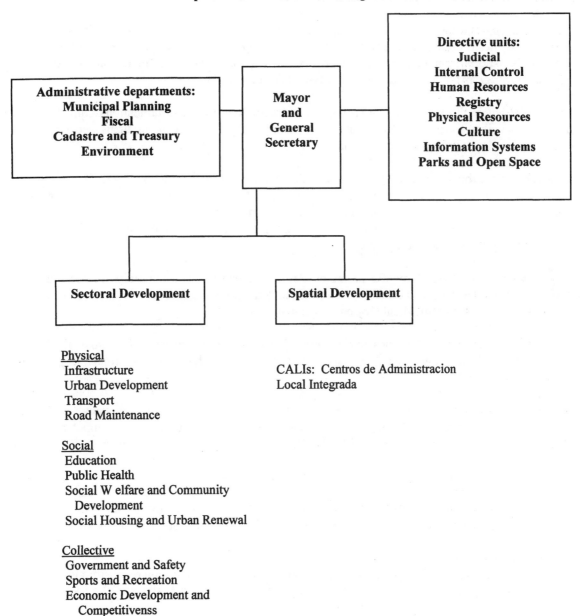

Source: Author's compilation, 1999.

• *Attention to specific groups.* This program provides training for disabled persons; strengthens programs to protect children, the elderly, and people in extreme poverty; develops and strengthens youth programs; and develops an information system to administer social programs.

• *Recreation, sports, and culture.* This program improves and maintains sports facilities, promotes a recognition of the importance of recreation and sports; supports various sports associations; develops comprehensive sports training programs; improves and maintains the infrastructure for cultural events; and promotes programs to disseminate cultural events throughout the city.

• *Justice and safety.* This program improves the infrastructure and communications equipment of the agencies responsible for public safety; promotes training of police; and strengthens decentralized and community-based conflict resolution mechanisms.

• *Housing.* This program restricts the construction of new housing in risk areas; strengthens self construction programs; improves land management; carries out urban renewal in the center of the city; acquires vacant land that is suitable for new development; promotes innovative alliances among communities, NGOs, and the private sector for the construction or improvement of housing; and maintains an inventory of irregular settlements.

• *Citizen participation.* This program promotes training; constructs and maintains public multiple-use facilities where citizens can gather; and develops an information system to keep citizens informed about the programs and processes in which they can participate.

• *Mass transit.* This program carries out a light railway, or metro, project with two main lines, one running south-north, and the other one east-west. A financial analysis of this project is presented in Chapter 7.

It is clear from the above that Cali's municipal development plan is a comprehensive program that proposes many necessary and feasible projects. In that respect, it is a very valuable input for the CDS. The obvious question is: if there is a municipal plan, why prepare a CDS? The answer is:

> - The plan has some shortcomings: (a) it proposes a multitude of activities without much selectivity and prioritization; (b) it does not include a financial analysis to realistically assess what can be implemented; and (c) it does not take into consideration the severe institutional crisis in the city at this time, which would make implementation of many of these projects difficult.

> - The CDS adds value to the municipal plan by engaging the city in a reconstruction process, by prioritizing the issues to be tackled, and by providing a sound analysis of the issues, including the financial situation of the municipality.

Other city programs

Cali has a multitude of interesting social programs that have helped to enrich the understanding of the CDS team. The development plan known as Municipio Saludable, for example, goes beyond health sector issues to include other social dimensions that make a city healthy. In addition, the municipal government's Social Policy Committee commissioned the NGO Foro Nacional por Colombiain to review social problems in Cali and propose new policies. The excellent report that resulted, "En Busca de la Equidad-Radiografia Social de Cali," has been extensively consulted by the CDS team. Finally, a large number of NGOs have developed exciting and sustainable programs, and the private sector and the

Chamber of Commerce have also been involved in supporting the city's social programs. These activities are, in fact, so extensive that the World Bank, through the CDS process, has commissioned the production of an institutional map of the city, to locate the main characteristics and areas of influence of the various social programs. This interesting and innovative work will enable policymakers to locate, quantify, and qualify the supply of social support in the city.

2. INSTITUTIONAL MODERNIZATION

Introduction

The story of Cali during the last five years is one of citizens' declining trust in the capacity of local government and business leaders to manage the city. The vacuum left by the decline in these institutions has not yet been filled. As a result, individuals and families pursue survival or quick-accumulation strategies, while the few businesses that have survived in the global environment have become islands with little or no connection to the city as a collective and dynamic organism. The absence of leadership has left the city without the means or the energy to mobilize local factors of production toward city-based growth and poverty alleviation.

Cali still has some strengths, however, on which basic social networks can be built. One is the widespread awareness of the present crisis and of the need to reconstruct governance. NGOs, businesses, universities, and churches are well aware that innovation is indispensable. Having failed with short-term, magic solutions, most city stakeholders seem prepared to contribute to medium or long-term processes geared to the reconstruction of social capital. They are also more open to learning from the good practices of other cities.

This chapter is organized as follows: First, the model of social capital accumulation that evolved during most of the twentieth century is discussed, as are the weaknesses and limitations that caused this model to collapse in the 1990s, when confronted with a combination of endogenous and exogenous forces. Second, the chapter discusses the challenges and demands that shape Cali's governance at present. And finally, the chapter proposes strategic priorities for institutional modernization in Cali.

Governance in Cali before 1990

This section summarizes the history of development and governance in Cali, as background for understanding Cali's organization and methods of operation (please refer to Table 2.1). The common thread that runs through the section is that of the slow development of a limited, largely informal (i.e., non-institutionalized), and highly vulnerable governance capacity that could not stand up to the challenges of the 1990s.

In the early 1900s, the city's economy centered on sugar production, based on an agriculture model in which large areas of land were cultivated with minimal labor input. As a result, <u>a few families</u> owned or controlled vast areas of one of the most fertile regions of the country. This was a key factor in determining power relations and organization of city government throughout the twentieth century.

In the period covering the 1910s through the 1930s, Cali was transformed into a commercial center by several important developments, notably the construction of railroads linking the city with neighboring towns and the port of Buenaventura. Development was further fueled by the official creation of the Valle del Cauca Department and the designation of Cali as its capital.

During the 1940s, the city's main economic activity shifted from commerce to manufacturing. Foods, textiles, paper, and publishing were the main industries at that time. Forestry, as well as sugar and cereal crops, were utilized as inputs for manufacturing. <u>Multinational corporations</u> invested in the area, often in close association with local economic interests. Frequent processes of vertical and horizontal integration further strengthened the family-based economic linkages between primary and secondary sectors within the area. State-regulated protectionism and state allocation of investment were also generally influenced or directly controlled by the same networks of local economic interests.

Table 2.1. Cali: city development and city governance, 1900-1990

Early 1900s
  Sugar cane plantations, factories.

1910s through 1930s
  Railroad linkage with the region, the country, and the seaport.
  The city becomes the road gate to the main Colombian seaport.
  With the above infrastructure developments, the city's economy revolves around commercial activities.
  Creation of the Valle del Cauca Department and designation of Cali as its capital.

1940s
  Construction of road networks connecting Cali with the rest of the country.
  Manufacture businesses consolidate and expand: foods, textiles, paper, publishing, cement.
  Foundation of the regional Universidad del Valle.

1950s and 1960s
  Creation of the city public utilities company as an independent entity.
  First public-private alliance, led by the Catholic church, for assimilation of first wave of internal immigration.
  Prominent business leaders receive training on social management, poverty alleviation, and urban development at the regional university.
  Annual city fair initiated.
  Strengthening of regional university.
  Water canals for irrigation and flood prevention.
  Development of a financial sector.

1970s
  Cali hosts the Pan-American Games. Road and service infrastructure developed accordingly.
  New regional airport.
  Closest interaction between business and city government. Business leaders serve as city and regional managers.
  Strengthening of city planning (primarily land use).
  Development of the valorization tax (following the precedent of Medellin).
  Cyclical ups and downs of property tax and valorization tax.
  Catholic church division over social issues and social policies.

1980s

Expansion of public-private alliances to recreation, environmental protection, micro-enterprise, productivity, and competitiveness.

The city experiments with client orientation and citizens' participation.

City government launches campaign for residents' identification with the city. The commandments of the caleñidad combine local idiosyncratic features with prescriptions for solidarity with fellow residents and with the city itself.

Public-private alliances prove successful in coping with second wave of massive internal immigration.

Creation and progress of NGOs oriented toward social programs, in particular Fundacion Carvajal.

Growing separation between regional university and business, city issues.

Late 1980s – 1990s

Gradual separation between local politics and business class. Decline of public-private alliances.

Growing number of NGOs primarily focused on city issues and on monitoring public administration.

Creation of several private universities, none of which yet has fulfilled the past role of the public regional university in linking business and city issues.

Further decline of regional university.

State decentralization makes evident the lack of hard budget constraint at the city level, and further exposes the city to the control of drug lords, patronage, aand corrupt politicians. In the absence of central government bailouts, the city is forced to austerity, and reduces current expenditures.

Globalization makes evident technological obsolescence and the need for restructuring of private corporations; and precipitates devaluation. Businesses going global struggle to escape from taxes, fees, and transactions costs that produce no corresponding benefits and reduce their global competitiveness.

Drug lords become influential in the city administration.

Growing intensity of internal armed conflict in the country. Police and military actions strengthen central government, in part by revising inter-government fiscal transfers, thereby adding to the city's fiscal pressures. Weakening legitimation of and trust in all levels of government authority.

Failed attempts to modernize city administration. The last administrative reform (1995-96) remains largely underdeveloped remains unfinished.

Deconcentration of city services fails to mobilize citizens' involvement or inspire decentralization of the city administration.

The combination of liberalization and globalization, internal armed conflict, and the war on drugs produces the deepest economic recession the city has ever experienced. The city suffers from high unemployment, rapid land devaluation, growing poverty and income inequality, high criminality, and violence.

Vanishing social cohesion and pursuit of individual survival or accumulation strategies.

Sources: Rojas, Ortiz, and Pinedo, 2000; Cámara de Comercio de Cali, 1992.

During the 1950s and 1960s, Cali witnessed many important changes. The rapid development of manufacturing in the city attracted a wave of immigrants. The first public-private partnership, between the city government and the Catholic Church, was established to accommodate the newcomers. Another public-private partnership, between the regional public university, Universidad del Valle and prominent business leaders, resulted in a the university offering a graduate program in business administration in which community and government leaders participated. The public sector was, in turn, strengthened by the growing leadership of the university, as well as by the creation of the a public utilities company and the completion of important infrastructure works.

During the 1970s, more infrastructure was built, particularly roads and tourism-oriented facilities in connection with the Pan-American Games. During this decade, collaboration between the business community and city government reached its peak. City planning, particularly land management, was strengthened. Public-controlled or public-financed investment in roads, railroads, canal irrigation, and airports substantially increased the value of land infrastructure. Inseparably tied to the allocation of public resources and largely dependent on the exploitation and valorization of land, powerful local economic interests found ways to mix and funnel public and private funds to poverty alleviation programs as well as to toward encouraging city residents to support urban development. City elites considered urban development and civic order as important factors in maintaining their power.

The 1980s brought new dynamics and actors. In the early part of the decade, the city government continued innovating, this time through new citizen participation models, citizen values, and a strong regionalism. Public-private partnerships, the church, and non-governmental organizations started paying close attention to the increasing social problems in the city due to drug trafficking (see para. 2.11) In 1986, Colombia implemented a major political reform, with mayors and city councilors elected by popular vote for periods of three years. Since then, Cali has had mayors who come from one of the two dominant parties, Liberal or Conservative. As in the rest of Colombia, the difference between the two parties gradually became blurry, and more and more people cast their votes based on the candidate's personal characteristics and proposals rather than on his or her political affiliation. The popular election of governors was instituted in the 1990s, as part of decentralization; and in Cali, in particular, interesting reforms took place. In each *comuna* was created a deconcentrated administrative office known as a Center of Integrated Local Administration (CALI). These centers are managed by the municipal central administration in consultation with the popularly elected Local Administrative Boards (JAL). They receive and control their own budgets, and they produce, with the input of the community, their own development plans, which feed the municipal development plan.

In the 1980s, another important phenomenon emerged: drug trafficking became the fastest and most popular way to accumulate wealth in Cali. Although the drug lords' profits came mainly from international markets—making them less dependent on land and locally based economic linkages—they followed the rent-seeking strategies that had long been in place in Cali—that, is, they sought control of the city and the regional government. In addition, they relied on land and land valorization for money laundering and political control.

Unlike past dominant economic interests, however, drug lords did not encourage residents to identify with the fate of the city. Rather, since their dominant position was based on their individual power and capacity to deliver services, drug lords sought loyalty to themselves as individuals. As a result of this shift in power, some local government officers and others tried to gain the confidence and support of the drug lords. But these new linkages could not be openly recognized. Thus the city underwent a shift from an informal but stable partnership between city elites and city government that openly controlled the city, to a situation of informal, uncertain, and rapidly changing partnerships that operated in relative secrecy.

Looking back on the city's development in the twentieth century, it is clear that one factor has always been central to the city's governance style: the culture of rent-seeking based on control of the city administration. Although prominent stakeholders and styles of government changed after drug activities came to dominate economic life, both the old and new types of goverance networks followed the same pattern of patronage and clientelistic politics and protection of monopoly (legal or illegal) profits based on the control of public budgets.

Before the 1990s, landowners, manufacturers, and merchants had organized themselves in such a way that rent-seeking activities were largely focused on public sector intervention in Cali and the surrounding Valle del Cauca Department. These rent-seeking strategies were, to a large extent, regulated by informal but effective pacts among city elites that identified their present well-being and their economic future with the fate of the city. That period was characterized by readily available leadership, creative public-private interaction, and considerable investment in the city's future. But underneath these positive aspects, what became evident during the 1990s was that a rent-seeking culture that is exclusionary in nature and operates via non-transparent incentives—no matter who controls it—cannot develop formal, open, and transparent institutions. That is, neither the overt leadership of traditional families nor the hidden leadership of drug lords would allow Cali to transform from a closed circle of tightly controlled political and economic interests into a political culture of formal and transparent rules and procedures.

Moreover, the particular type of social capital that had been carefully crafted over 70 or 80 years of land-based economic development proved to be fragile and ephemeral, and contained the seed of its own destruction. The limited depth and coverage of participatory institutions (not sufficiently rooted in civil society and largely restricted to a few traditional families) made civil society dependent on the leadership of traditional elites, which failed to provide equal opportunities in city management when the city grew exponentially and new demands and centers of urban power began to appear. When the traditional elites were challenged—and to some extent defeated—by the new emerging powers associated with drug trafficking, the whole model of governance and social capital accumulation collapsed. Graph 2.1. summarizes the rent-seeking model of governance in Cali.

The challenges of governance in present-day Cali

As a result of these conditions, there has been a growing distrust between local government and business and civil society organizations since the mid 1990s. This can be seen, for example, by the inability of either public or private sector agents to mobilize other local actors for economic initiatives or poverty alleviation. On the contrary, the economic stagnation that began with the war on drugs of the mid 1990s has been reinforced by atomization of the city's resources; local public policies suffer from discontinuity between one administration and the next; and in any case, local govenrment lacks the fiscal and human resources needed to execute many of its policies.[4] At present, no formal or informal institution appears capable of re-establishing the trust and leadership that are needed to rebuild the city in the context of a globalized economy.

---

[4] Discontinuities in the reform of city government have been evident since the late 1980s. For instance: (a) the recommendations of a widely participatory and carefully crafted city pact known as El Cali que Queremos were never implemented; (b) both the public and private sectors of the city paid for a Monitor study for proposals to enhance city competitiveness, but the recommendations in that study, too, went largely unapplied; and (3) the Administrative Reform Act of 1995-96, based on in-depth studies prepared by the Universidad del Valle's School of Business Administration, was not implemented by the administration that took office in 1998.

Graph 2.1. Rent-seeking model in Cali
Twentieth century

BEFORE 1990

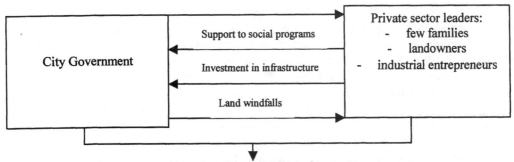

- Investment in city's future
- High concentration of political and economic power
- No institutionalization of public-private partnerships

AFTER 1990

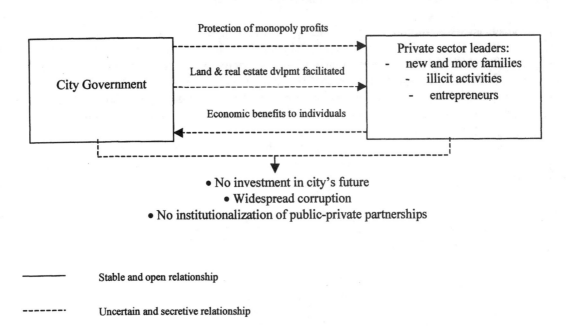

- No investment in city's future
- Widespread corruption
- No institutionalization of public-private partnerships

———— Stable and open relationship

-------- Uncertain and secretive relationship

Source: Author's compilation, 1999.

Three exogenous developments have further debilitated the capacity of the city to establish a credible and effective government. *First,* liberalization policies were adopted by the central government during the early 1990s, to foster economic restructuring and global competitiveness, precisely at the peak of drug trafficking in Cali, which meant the city was ill-prepared for these initiatives. *Second,* the state decentralization policies, which began in 1986, made cities more dependent on their own resources. As Cali has learned, fiscal decentralization means hard budget constraints, fiscal responsibility, and no government bailouts. When the city's financial crisis became evident in 1998, the central government denied the rescue requests of local political leaders. Then international creditors imposed a fiscal austerity program on the city, which further debilitated the patronage-clientelist networks that supply most local government staff, and ignited trade union protests at a time the city government was least prepared to mobilize either resources or public opinion. *Third,* the country's armed conflict has intensified during the last ten years, requiring more centralized command of the police, the army, and public resources, as well as an increase in the national tax rate, thus weakening local governance and crowding out local tax initiatives. At the same time, the armed conflict has made evident government and party weaknesses, thus raising doubts about the legitimacy of the national government.

But there is hope. The crisis of governance and city management has become an opportunity for new leadership and management innovation. In this section, the profiles of some of the potential participants in constructing Cali's governance are examined, in no particular order.

Cali has always been a civic-minded city, with its <u>citizens</u> very much involved in improving urban life. In spite of enormous risks and difficulties, groups of citizens have been fighting corruption and crime with passion and conviction. Caleños have a great desire to be actively engaged in city affairs. Almost a quarter of the population, according to the recent Household Survey to Measure Access and Satisfaction with the Municipal Services of Cali (Encuesta de Acceso, Percepción de los Servicios Ofrecidos por el Municipio de Cali, EPSOC) prepared for the Cali CDS (please refer to Chapter 4), would be interested in participating in the organization, reform, or administration of education, health, and employment programs. And more than 10 percent said they would actively work with the police, despite heir ambivalence toward them. This potential for civic engagement is important – and it spans to the political realm as well.

Graph 2.2. Willingness of the population to participate in government programs

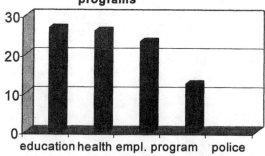

Source: ESPOC, 1999.

Unfortunately, given the tremendous city crisis, real civic engagement in public and social life has decreased. Overall, according to EPSOC, the participation rate in certain community organizations (the choices were *Juntas de Accion Comunal*, religious groups, women's groups) stands at 8 percent in the city, with some variation among income groups (Table 2.2). Of those who are participating, religious and women's groups are relatively more popular among the poor, while *Juntas de Accion Comunal* are more popular with higher-income groups. And the primary factor limiting such participation—espcially for the poor—is not time, as one might expect, but the fact that existing organizations do not meet their expectations and demands. It is important to note that the concept of popular participation is broader than these three options; it includes other activities such as peace marches, safety networks, sports events, etc. The purpose of the EPSOC survey, however, was to illustrate the level of participation through established political channels (the JACs), versus more grassroots / community-based options (the religious and women's groups).

Table 2.2. Cali, participation rates in community affairs, September 1999

Community affairs	Income quintile					Total
	1	2	3	4	5	
Overall participation rate in organizations	8.3	7.1	9.4	6.9	9.6	8.3
If participating:						
- JAC	16.4	24.0	11.9	26.4	29.8	21.5
- religious groups	35.3	17.5	15.0	23.3	15.0	20.9
- women's groups	25.9	30.6	44.2	27.3	12.7	28.0
If not participating:						
- no time	32.3	37.6	40.8	36.4	49.1	39.2
- organization not liked	40.7	34.9	35.7	38.4	27.1	35.4
Junta de Accion Local:						
- member known?	39.4	42.0	42.6	39.9	35.5	39.9
- good mechanisms?	76.6	74.9	77.7	79.7	71.7	76.1

Source: EPSOC, 1999.

In spite of growing difficulties, the <u>municipal government</u> continues to be a main actor in Cali, leading many social and infrastructure projects, preparing the formulation of a comprehensive social policy, and strengthening a decentralized system of city management. But it faces many problems, among them: (a) a serious financial crisis; (b) an over-staffed and inefficient human resource system; (c) lack of credibility among constituents; and (d) involvement in too many disperse and often overlapping social programs. Points (a) through (c) are thoroughly covered in Chapter 7, and point (d) is discussed at length in chapters 4, 5, and 6. Attempts of the present administration to reduce staff expenditures have failed, generating protests from municipal labor unions and the public utility companies. To overcome these difficulties, the municipal government has no choice but to associate with other city stakeholders in developing a new vision for the city. Details on this are presented in the last section.

As noted above, <u>churches</u> (Catholic and others) have enjoyed considerable leadership in Cali, particularly in the design and implementation of social programs that work toward poverty alleviation and equitable urban growth. The EPSOC survey found that 21 percent of households participate actively in religious groups (see Table 2.2)—a percentage almost as high as for participation in established political participation channels (the JACs). When the numbers are broken by income quintile, the results show that religious groups play an even more important role among the poorest, with a 35 percent participation rate in religious groups among households in the first income quintile. In Aguablanca, a poor section of Cali, for instance, there are three large religious organizations that operate social programs: (a) the parish organization *El Señor de los Milagros*, directed by Father Welker, has two schools for primary and secondary education, covering 11,000 students, a day care center for 1,000 children, and a health center, all offering services at very low cost; (b) the program *La Casita de la Vida*, directed by Sister Alba Esthella, works to prevent teen pregnancies and provides medical services to pregnant women; and (c) the program *Sol y Vida*, run by Cali's archdiocese, is dedicated to reducing violence and strengthening peaceful coexistence (see Chapter 5 for more details on particular program).[5] The priests and nuns who lead these programs are broadly perceived as promoting civic involvement. It is important to note, however, that the church leaders involved in these basic social networks do not act primarily as representatives of a particular religious denomination, but on the basis of their individual leadership capacity, and that they are trusted because they work for the people rather than through the political system.

Cali has had an important tradition of active NGOs. As of March 1999, there were 400 registered NGOs, of which 230 were affiliated with the association of NGOs, PROCALI. Of these 230, 115 are considered small, with assets valued between $2,500 and $10,500; 107 medium, with assets of $10,600 to $50,000; and 8 large, with assets of more than $50,000. These NGOs work in a wide range of sectors, covering social, environmental, institutional, economic, and housing programs (please refer to Table 2.3.). In addition, PROCALI maintains close relationships with: (a) the municipal government, with which member NGOs participated in 293 projects valued at more than US$3 million dollars in 1998 (please refer to Table 2.4). PROCALI also maintains relations with the Spatial Planning Council, the Social Committee, and the Healthy Municipality Program, and participates in policy negotiations related to youth and family; (b) the private sector, where PROCALI participates in the Permanent Entrepreneurial Committee of the Valle del Cauca Department (*Comité Empresarial Permanente del Valle del Cauca*), the Leadership and Solidarity Front in the Valle del Cauca Department (*Frente de Liderazgo y Solidaridad del Valle del Cauca*), and the Chamber of Commerce; (c) churches, with many NGOs working with Cali's archdiocese and other religious groups; and (iv) academia, particularly through internship programs in the Universidad Javeriana, and on particular topics with four of the most prominent universities in the city.

NGOs in Cali face a series of difficulties: (a) although they work hard to carry out their altruistic mission, they are seldom managed efficiently as businesses, resulting in program cutbacks and lack of continuity; (b) they need more negotiation capacity to work effectively with government at all levels, while maintaining their independence; (c) they do not have effective evaluation systems; (d) they tend to act independently of one another; and (v) they need to try more innovative approaches to solving these and other problems. PROCALI, created in 1980, has been instrumental in solving some of these problems, and particularly in shifting the role of NGOs from isolated assistance to communities and individuals to suppliers of strong social programs, and from short-term impact to longer-term public relevance. However, NGOs have not escaped the present economic crisis in Cali, and have had to reduce their staff, cut programs, and establish creative partnerships in order to survive.

As for the city's entrepreneurs, the economic conditions in the country and in the city have changed the rules of the game. The business world is witnessing important changes: (a) the entrance of economic groups from other regions of the country, the most notable case being the participation of the *Ardila Lulle* group (from Medellin) in sugar production in Cali. a sector that had always been dominated by traditional families from Cali; (b) the creation of strategic alliances between local entrepreneurs and multinationals, two examples being the privatization of solid waste collection, and the sale of a well-established Cali company, *Varela*, which produced personal hygiene articles; (c) innovations in production processes, administration, and product distribution in companies that are trying to remain competitive; and finally

---

[5] This information has been captured through the institutional mapping carried out by the Socioeconomic Research Center at Universidad del Valle (CIDSE) and financed by the World Bank under the Cali CDS.

(d) diversification has become an important strategy in large traditional companies as a way to protect business from changing economic conditions, an example being *Manuelita SA*, a sugar producer, which is now involved in textiles, shrimp production, ice production, and even real estate.[6] These changes suggest that traditional business leaders in Cali cannot play prominent roles as facilitators of the proposed social networks for two reasons: (a) they have lost economic leadership in the city at the expense of external groups; and (b) some of them may be seen as beneficiaries of pre-existing models of capital accumulation. Due to the loss of stature of these business leaders, the Chamber of Commerce has had to fulfill the facilitator role, funneling resources from the private sector (both traditional and new companies) to civic and social programs. What lies ahead is for the Chamber of Commerce to channel support for the reconstruction of Cali, particularly through education and youth programs.

Table 2.3. NGOs affiliated with PROCALI, by sector

Sector	Number of NGOs
Education and citizen participation	58
Health and rehabilitation	48
Childhood and youth	38
Ecology and environment	31
Women and family	27
Community development and voluntary work	27
Culture, recreation and sports	24
Institutional development	19
Economy and employment	16
Senior citizenry	12
Housing and urban development	6
Security and emergencies	5
TOTAL	311 (An NGO can work in more than one sector)

Source: Pinedo 2000.

Academia is another important stakeholder in Cali, and both individual professors and researchers and institutions should be encouraged to participate in city affairs. A consortium of universities and research institutions, for example, could organize public discussions on restructuring of local businesses and public corporations, evaluate city government and private sector collaboration, and offer specialized courses on urban studies and training sessions on governance.

Other actors will be important in rebuilding Cali: (a) genuine community leaders (including leaders of new urban communities made up of people displaced from rural areas by the ongoing armed conflict), whose reference groups are the communities themselves rather than party structures or patronage networks; these leaders will need incentives to prevent co-option by the patronage networks of political parties;[7] (b) local mass media and public opinion leaders whose influence is independent of the political parties; (c) the police, in spite of multiple difficulties and high risks, because they have been active in fighting corruption, drug dealing, and common crime, and—after being restructured nationally- have

---

[6] For more details on the changing business environment, please refer to Urrea, n.d.
[7] The Cali experience with the officially sanctioned community action boards, known as JALs and JACs, has been one of co-optation by political parties and integration of their leaders within larger patronage networks.

renewed their commitment to building social cohesion through crime prevention and citizens' participation; (d) former guerrilla members, who—as exemplified by the ongoing experience of the neighboring municipality of Yumbo[8]—can effectively play the role of honest brokers between business and community leaders, and can also command the respect of present guerrilla groups, the armed forces, and defenders of human rights.

Table 2.4. Projects jointly executed by NGOs and the municipal government, 1998

Secretariat/Department	No. of projects	Value (in USD $)
Social welfare and community development	99	879,233
Health	62	270,115
Culture	43	162,550
Education	36	1,684,373
Government and safety	24	170,559
Sports	10	51,000
Spatial development	9	56,182
Economic development and competitiveness	6	51,096
Environment	4	47,000
TOTAL	293	3,372,108

Source: Pinedo 2000.

Proposed strategic priorities for institutional modernization

The current challenge for Cali is to adapt to, and take advantage of, the new demands of the globalized economy in the context of increasing local autonomy. A transparent and efficient local government that responds primarily to citizens' priorities and pursues the creation of an adequate competitive environment is a *sine qua non* for the city to become part of the dynamic new international economy. However, the city's long-entrenched tradition of patronage and rents, based on the control of land and local government, cannot be changed overnight. Entrenched interests in current city management can be expected to pose formidable resistance. New incentives have to be realigned time and again to stimulate mutual trust and win-win public goods projects. Medium and long-term goals have to be based on short-term results and demonstration effects.

Reconstruction of the city's most basic social and political networks has been identified by the CDS teams as the most immediate priority for Cali's governance and urban development. The reconstruction should be facilitated by actors who can reconcile their own interests with those of the city. These basic networks should be inclusionary, expanding themselves through ever-growing circles. The transparency of each participant's interest is a prerequisite for trust-building within these networks, which should work toward achieving modest, intermediate goals that stimulate self-confidence and motivate pursuit of more ambitious goals. In addition, the city should undertake the following initiatives along with city stakeholders and, wherever possible, the Department (regional government) and associations of neighboring municipalities:

---

[8] Yumbo is the main locus of large manufacturing activity within the Cali metropolitan area. The present mayor, Rosenberg Pabon, is a former leader of a guerrilla group that demobilized in 1990 and subsequently transformed into a legal political movement. Mr. Pabon was recently voted the most popular mayor of Colombia's 1,040 municipalities.

• *Carry out a serious review and evaluation of past governance initiatives.* First, the city government should evaluate the impact and sustainability of previous public-private alliances and recommend design and incentives for new alliances. Second, the city should review the last ten years of studies and proposals for city reform (Program *El Cali que Queremos,* Monitor Study, Administrative Reform[9]), to identify the reasons for lack of implementation; and should widely disseminate the review through low-cost means (i.e., the media).

• *Promote and implement public forums to discuss the future of the city.* The city government, along with city stakeholders, should promote public forums in order to: (a) discuss the city's main problems, using (and thereby endorsing) the findings of this report, and perhaps publish a summary version of the report in local newspapers; (b) achieve consensus about a limited number of specific short-term projects, based on the report's proposals; (c) reach agreement on measures to reform city government and reduce unnecessary staff, and on the design of incentives to create a culture of efficiency and responsiveness to clients; and (d) develop a system of civil society sanctions against corruption.

• *Listen to potential city investors.* The city government and stakeholders should invite national and international investors to identify their concerns and expectations regarding city services, city infrastructure, and adequate investment conditions.

• *Seek technical assistance in city management* . The city government and stakeholders could promote exchanges with and technical assistance from other cities that have developed good practices with regard to citizens' participation in city management and control.

• *Seek the support of external actors.* Since the lack of transparency in individual actions and city management has created a climate of distrust, external facilitators and incentives are needed to overcome apathy, mutual blaming, and pessimism. The following guidelines may help ensure effectiveness in external support:

- The contribution of external agents and incentives should be conditioned on—and measured by—growing local ownership and sustainability of the new city networks.
- More than financial incentives, what a potentially self-sustaining city such as Cali seems to need most is the stimulus of credible, creative ideas, and other cities' success stories.
- Small-scale pilots or demonstration experiments within the city may also awaken the dormant creativity of the Caleños.
- The coordination of cooperation agencies in conveying strong, consistent signals about the need for reform is required to transform Cali's bureaucratic culture and overcome citizens' apathy.
- Coordination with the central government's urban and decentralization policies is also desirable.

In sum, external agents should be given a limited, catalytic role in revitalizing and mobilizing the largely dormant strengths the city still possesses.

---

[9] Cámara de Comercio, 1992, Cámara de Comercio, n.d.

ECONOMIC REACTIVATION[10]

Introduction

Cali is in the midst of what may be the worst economic crisis in its recorded history, with a GDP that has fallen since 1995 by more than 2 percent a year. At the same time, the country is also in severe recession, and the city cannot, in the short term, separate itself from this wider context, including the effects of the national trade regime, fiscal and monetary policy, and exchange rate policy. Thus, despite the urgent need for reform, Cali's ability to sustain a recovery is, in many ways, constrained by factors beyond the city's control.

There are important practical constraints to the economic reactivation component of the Cali CDS. First, a city is normally defined by an administrative area, but in economic terms this area is entirely arbitrary; that is, Cali as an economy cannot be easily separated from the Valle del Cauca Department, the Paez region, and many other localities whose contributions make it possible for Caleños to earn a living. Thus, the definition of the city's economic output is inevitably imprecise. Second, there are major economic data deficiencies regarding Cali. Data about the informal economy are sketchy and anecdotal, although this sector might be an important source of dynamic growth and change as well as a safety net for those who have recently lost jobs in the formal economy. At the same time, national statistical systems are ill-equipped to collect the most useful data at the city level. Often it is difficult to separate reliably the city from the Department, much less identify in the city the groupings of specialized economic activity that are vital for understanding patterns of spatial interaction. Similarly, the categories used in presenting national statistics are far too aggregated to be useful at the city level.

This chapter is organized in seven sections. The next section provides an overview of Cali's economy, including basic data and context. The third section describes changes in Cali's labor market in the 1990s. The fourth details the outputs and employment figures of the main economic sectors. The fifth section analyzes exports as one of the city's comparative advantages and a possible key element in the city's economic future. The sixth section discusses social and physical infrastructure as facilitators of the city's economy. The final section proposes a program of actions based on the analysis of all of these elements.[11]

Overview of Cali's economy

To gain an understanding of Cali's economy, it is important to examine the national context over the past decade, as well as some departmental statistics and the dynamics of Cali's main economic sectors.

The Colombian economy expanded at a moderate rate in the first half of the 1990s, faltered in 1996-97, and then moved into severe recession. The components of the recession, as enumerated by informed commentators, include (in no order of priority): (a) the drawn-out effects of an economic opening that was not accompanied by significant improvements in infrastructure and labor qualifications, which are essential for competitiveness; (b) important changes in government policy to control the fiscal deficit and high rate of inflation, including increased taxation, high rates of interest, a freeze on public expenditure, and cuts in public employment; (c) changes in external prices (coffee, petroleum), which were important contributors to the balance of trade deficit; (d) currency changes and increased risks as the result of the

---

[10] At the time this report was being revised for publication, much more information was collected by a new economic observatory in Cali, created at the recommendation of the World Bank. Some of the numbers and tables in this report have been updated to reflect this new information. The economic trends described in the chapter, however, remain unchanged, with only some signs of recovery of the Colombian economy that might have spillover effects in Cali.

[11] The local economic team has continued to gather information on Cali's economic activities after this report was finalized. The report of this team will be an extraordinarily valuable source of information for the next stages of the CDS process.

financial crisis in East and Southeast Asia, which were reflected in Russia and Brazil and affected the inflow of foreign investment and thus the capital account; and (e) a severe financial crisis (from the third quarter of 1998) and the disruption of credit markets, which exaggerated problem of high interest rates. By the time interest rates were reduced, general recession had radically cut the demand for capital, so that devaluation (September 1998) increased the cost of imports (with negative effects on the inputs to domestic production) without raising incentives to export (credit growth declined by 29 percent in 1997, 12 percent in 1998, and 6 percent in 1999). In sum, a decline in demand became the major factor in the persistence of recession, which was exacerbated by the continuing effects of the disruption of the narcotics trade (the Cali Cartel was eliminated in 1995) and increasing domestic warfare.

The Valle del Cauca Department has not escaped Colombia's economic recession. Although the department generated, in 1995, 13 percent of Colombia's GDP,[12] and has consistently, since 1980s, been the country's third largest departmental economy (after Cundinamarca and Antioquia), Valle del Cauca is now facing the worst economic recession in recent history. Graph 3.1 shows the real growth rate of GDP for both the country and the Valle del Cauca Department. Four periods can be identified: from 1981 to 1986 the department's rate of GDP growth follows the national trend; the period 1986-1991 is characterized by fluctuations; from 1991 to 1995, the department's GDP grew much more than that of the country; but from 1995 on the trend is reversed.[13]

In terms of economic sectors in the Valle del Cauca Department, these have consistently ranked as follows since 1980: industry (contributes 15 percent of the national industrial GDP), agriculture (11 percent of the national agricultural GDP), commerce (12 percent of the national commercial GDP), and finance (12 percent of the national financial GDP).[14]

Graph 3.1. Real growth rate of GDP: Valle del Cauca vs. Colombia

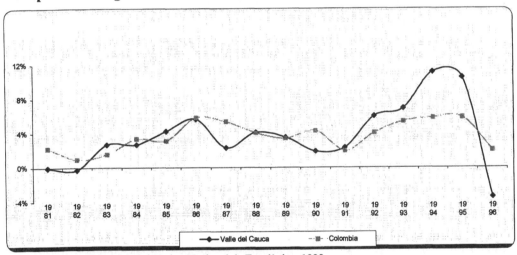

Source: Departamento Administrativo Nacional de Estadística, 1999.

Colombia's National Statistics Department (DANE) reports GDP figures for the country and for the departments but not for cities. The Municipality's economic team has calculated GDP figures for Cali using DANE's methodology. Although these numbers should be taken with caution, as this is the first

---

[12] National Statistical Department (DANE).
[13] More recent data show that GDP has continued to decrease steadily both in Colombia and in the Valle del Cauca Department.
[14] Mision Siglo XXI, 1999.

time they have been calucated, they are useful indicators of main proportions and trends. According to these calculations, Cali's GDP was about US$6 billion in 1996, representing 7 percent of the national GDP. Graph 3.2 shows that the main sectors in terms of Cali's GDP share are: manufacture, personal/social/communal services, real estate services, financial services, construction, commerce, and agriculture. All of these except financial services have had negative growth since 1996.

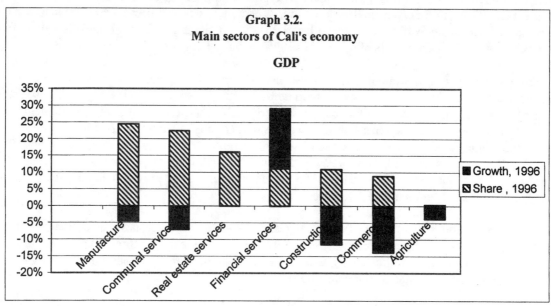

Source: Secretaría de Fomento Económico y Competitividad, 1998.

Cali's labor market

Cali experienced a remarkable 18.6 percent growth in overall employment until 1998, but this growth varied significantly among sectors (please refer to Table 3.1). Mining, for example, declined dramatically (-31 percent), and many other sectors declined in relative terms between 1990/91 and 1997/98—especially construction[15] and industry, which continued to suffer from the de-industrialization trend. Commerce and personal, social, and communal services, also showed marginal relative declines. The growth sectors were transport (55 percent) and financial services (87 percent).

These sectoral changes were accompanied by changes in the occupational structure (please refer to Table 3.2). There was disproportionate growth in the percentage of self-employed (from a quarter of the recorded labor force at the beginning of the decade to a third in 1998). The only other sector to grow was "unpaid family workers" (by a third), but the share of these workers in official employment is small (1.5 percent in 1998). In almost all other categories, there were relative declines, the highest being in private employment. The public sector, both employees and employers, had an absolute decline.

Although Cali performed well, until 1998, in the creation of jobs in the official economy, the growth of the labor force was even greater—the rate of participation, 58 percent in the fourth quarter of 1996 (compared to 51.8 percent in 1976), increased to 65 percent in the fourth quarter of 1998. This may have reflected the arrival of minors at working age, increases in immigration, and higher rates of participation

---

[15] Construction is a cyclical sector, so figures for that sector may reflect a downturn in the cycle related to the recession, rather than a trend.

by those who had been inactive (the aged, children, housewives). The conjunction of these factors resulted in a rapid increase in the official figures for unemployment, which more than doubled in the five years up to 1998. Between 1998 and 1999 alone, employment of skilled and unskilled workers declined by 35,000. In terms of sectors (please refer to Table 3.3), the key sources of contraction have been construction and trade/commerce, followed by finance, industry, services, and transport.

Table 3.1. Structure of employment in Cali, 1990/94 and 1995/98

Sector	1990/94 average % share	1995/98 average % share	% change
Agriculture	1.0	0.9	+2
Mining	0.2	0.1	-31
Industry	24.4	21.6	+5
Electricity	0.7	0.6	+8
Construction	6.6	5.8	+4
Commerce	26.8	26.6	+18
Transport	5.2	6.9	+55
Financial services	5.6	8.8	+87
Personal, social, communal services	29.4	28.4	+14
Total	100.0	100.0	+18.6

Source: Mision Siglo XXI, 1999.

Table 3.2. Occupational status of the labor force in Cali, 1990/94 and 1995/98

Status	1990-94 average share (%)	1995-98 average share (%)	Change (%)
Unpaid family worker	1.4	1.5	+33.7
Private employee	55.0	49.7	+7.2
Public employee	8.1	6.5	-5.2
Domestic labor	5.7	4.9	+1.1
Self-employed	24.2	33.0	+61.6
Employer	5.6	4.4	-7.3
Total	100.0	100.0	+18.5

Source: Mision Siglo XXI, 1999.

Table 3.3. Unemployment in Cali by sector, 1997

Sectors	Percent unemployed
Industry	18.9
Construction	25.5
Transport	15.9
Finance	22.9
Trade, commerce	25.5
Services	16.5

Source: Camara de Comercio de Cali, 1999.

The structural changes in employment have been unequally distributed: the poorest *comunas* now have up to 30 percent unemployment, while the richer ones have less than 10 percent (please refer to Table 3.4).

Contrary to other data sources, the EPSOC[16] (household survey) found male unemployment to be somewhat higher than female unemployment (18 versus 16 percent). Such aggregates hide significant variation across age groups. For example, young women 14 to 24 years of age have the highest unemployment rate of all groups, reaching 56 percent in the lowest income group. Overall, youth unemployment is much higher than that of older age groups: the rate for 14 to 24 year olds is twice that for 25 to 39 year olds, and three times that for people aged 40 and above. The rate for those who have completed secondary education is higher than for those with only primary or tertiary education (25.1 versus 17.8 and 7.9 percent, respectively). This means that Cali's labor market is a world of low-paid workers (*peones*) and professionals (*doctores*), with few in the intermediate categories.[17] For more details refer to Chapter 6.

The rise in unemployment would have been even higher had the informal sector not absorbed a large number of workers formerly employed in the formal sector. At the same time that worker employment was declining, self-employment increased strongly, by about 28,000—almost absorbing the decline. Such increase in informality, however, goes hand in hand with an increasing share of the labor force without work contracts or access to social security. As Table 3.4 shows, more than 70 percent of the employed in the poorest income group do not have formal work contracts, and only 13 percent have access to social security.

In addition, several other characteristics of Cali's labor market stand out. First, the duration of unemployment has increased from 5 months in 1994 to 8 months in 1998.[18] Second, with increasing unemployment has come increasing under-employment (i.e, people work less than full time). Third, all factors being equal (i.e., experience and years of education), women earn less than their male counterparts in Cali—gender discrimination in the labor market exists.[19] And fourth, a certain *area effect* plays an important role in determining wages: if workers come from all areas except the central corridor (*comunas* 2, 17, 19), they tend to earn comparably less, although they have the same background, years of education, background, etc. This could be a stigma or social class effect, or it might reflect differences in the quality of education across different *comunas*. In any event, such inequities in pay contribute to inequalities within the city's population.

Economic sectors

This section discusses three main economic sectors: (a) manufacture; (b) traded services, including commerce, finance, and personal/social/communal services; and (c) construction. It also discusses informal employment, which cuts across all sectors. No data are available, at this time, on real estate services, even though that sector represents almost 16 percent of Cali's GDP. Data on informal employment are based on the results of the national household surveys, which are, at present, the best means for estimating the informal economy.

<u>Manufacturing</u> boomed in the city—as in all of Colombia—as a by-product of the Great Depression of the early 1930s and the resulting enforced limits on imports. Supplying the domestic market sustained high rates of growth—up to 11 percent annually—until the 1970s, with the peak phase of growth occurring between 1930 and 1964. This period included an acceleration in the 1950s due to the discovery of local energy sources and rising world demand for Colombia's raw material exports. As a proportion of Cali's

---

[16] Household survey prepared for the Cali CDS. A total of 1,912 households were included in the sample. Please refer to Chapter 4 for more information.
[17] Urrea and Ortiz, 1999.
[18] Urrea and Ortiz, 1999.
[19] A Minceranian equation was estimated with the log of hourly wages regressed against individual and household characteristics (education, age, gender, type of occupation, area of residence, etc.). The regression is reproduced in Appendix 1 of Chapter 4.

growing labor force, manufacturing moved from 28.5 percent at the end of the First World War, to 35 percent in 1938, to 37.5 percent in 1951, to 29 percent in the 1970s.

Table 3.4. Characteristics of the labor market in Cali, September 1999

Characteristic	Income quintile					Total
	1	2	3	4	5	
Unemployment rate	35.9	22.4	18.4	11.8	5.8	17.1
- male	35.7	24.6	18.4	12.9	6.4	18.0
- female	36.1	18.8	18.5	10.0	5.0	15.9
By age						
14-24 yrs	51.8	34.8	27.7	21.7	19.7	30.3
- male	48.5	34.3	29.3	25.5	23.3	31.6
- female	56.0	35.8	25.7	16.2	15.1	28.3
25-39 yrs	30.6	15.4	19.8	10.8	3.5	14.6
40 and above	28.6	21.4	8.1	4.4	1.7	10.2
Labor market participation rate	51.9	54.7	60.3	63.9	65.7	59.8
- male	78.5	78.5	79.1	81.1	76.8	78.9
- female	35.2	36.4	44.9	47.8	55.4	44.2
Employment						
- w/o contract	71.2	51.7	46.8	39.5	26.7	42.3
- w/o access to pensions	15.6	34.5	32.2	42.0	56.2	40.2

Source: EPSOC, 1999.

The former protectionist regime in Colombia up to the 1990s sustained relatively high rates of growth, promoting in Cali a group of large national companies in engineering, food processing, paper and publishing, chemicals, and cosmetics. Many multinational companies in the fields of food processing and chemicals, notably pharmaceuticals, located in Cali in the 1960s, but direct foreign investment stagnated beginning in the mid-1970s, and there have been no new foreign-owned enterprises opening branches in Cali since then.

The manufacturing sector in the greater Cali area is a mixture of high-growth, labor-intensive, export-oriented light industry (food processing, paper and packaging, garments, and shoes), and capital-intensive heavy and medium industry (chemicals, metals, synthetic materials), focused mainly on the domestic market. Table 3.5 contains data on the share of value added of the four most important industries in Cali before and after the opening of the Colombian economy in the 1990s. The structure has not changed dramatically, with chemicals and foods experiencing relative increases in importance, and paper and publishing maintaining their relative position. On the basis of Colombia's income level, it seems unlikely that the country will, in the future, become an important supplier in the labor-intensive fields, since it would have to compete with much lower income level countries, but perhaps the selection of items produced (unbranded components of garments, for example, rather than finished products) would enable Cali to be competitive.

The manufacturing sector officially employs more than 80,000 people. Of these, the chemical industry employs 26 percent, the textile and leather industries employ another 26 percent, the food processing industry 16 percent, and paper and publishing 13 percent.

In the 1990s, the manufacture share of Cali's GDP saw an overall decline (please refer to Graph 3.3). This was the result of a general trend of de-industrialization in the city and a deconcentration of industrial locations in surrounding municipalities, notably Yumbo to the north, and Puerto Tejada, Santander de Quilichao, and Caloto, to the south in the neighboring Cauca Department.

Table 3.5. Cali's industrial structure, 1980-84 and 1995-97

Industry type	Share of industrial value added (%) 1980-84	Share of industrial value added (%) 1995-97
Chemicals (pharmaceuticals, cleaning products, painting products)	22	28
Foods (fats, oil, sugar, sweets)	8	14
Paper	13	11
Publishing	6	7
Textiles and leather	8	5

Source: Mision Siglo XXI, 1999.

Finally, in terms of geographic location, industrial activity within the city has been concentrated around the downtown area, in the northeast of the city (please refer to Annex A). Up to 1990, *comunas* 18, 17, 10, 11, and 12, to the south, had significant levels of industrial employment, in addition to *comunas* 3, 4, 5, 6, 7, 8, and 9 to the northeast. By 1999, industrial activity was clearly more concentrated in four *comunas:* 3, 4, 8, and 9, which surround the downtown area, suggesting that these are light industries that profit from the services and commerce activities of the core. Medium and heavy industry has increasingly moved toward Yumbo in the north.

Cali's industries have a high level of energy consumption per unit of output because of the composition of output—steel, cement, paper, metals, and chemicals. Energy supply has been a problem in the past, although it is now said that new investment will relieve this bottleneck. However, such industries—with their frequently negative environmental impacts—perhaps represent the past and are likely to be phased out as the unit value of output rises.

In the case of traded services, the data are much more limited, even though this group accounts for two thirds of the value added in the official city economy, and 58 percent of Cali's GDP in 1996.

Commercial activities represented almost 9 percent of Cali's GDP in 1996, with a continuous decreasing trend since the beginning of the 1990s (please refer to Table 3.1). The sector employs more than 25 percent of the labor force, and this indicator, in contrast with the previous one, has remained relatively constant throughout the 1990s. The largest share of establishments (15 percent) were in the supply of groceries, 87 percent of the total which were micro-enterprises, including – in the spring of 1996 – some 3,575 wandering street sellers. About a third of the micro- enterprises are estimated to last less than one year. There has been a significant decline in the number of micro establishments in the past three years, but a growth in the number of street traders. However, these overall figures do not allow us to see the long-term shifts in types of trade (the emergence of growth or declining sectors), nor the different markets (local consumption versus purchases by visitors to the city). Nor do the figures show patterns of specialization by locality within the city, where public intervention to facilitate growth might be useful (see below on wholesale markets). It is also necessary to separate out restaurants and hotels as facilities particularly related to the potential for expanded tourism.

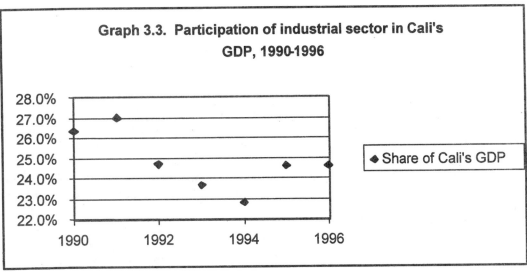

Source: Mision Siglo XXI, 1999.

The financial GDP of Cali has represented, since the 1980s, between 10 and 14 percent of the national financial GDP, with a decrease in the period 1985-89 and an overall increase in the 1990s. Financial activities also represent 11 percent of Cali's own GDP, and comprise the only economic sector that exhibited growth in 1996. The sector employed close to 10 percent of the work force in the late 1990s—several percentage points above the corresponding figure in the early 1990s (please refer to Table 3.1.). Concentration of financial activity in Cali is also measured by the following indicators: 298 offices of commercial banks, 89 of savings and housing finance, 40 of financial corporations and companies, and a stock exchange.

At the time of writing, information on the scale, composition, and specialization of the health sector is not complete, though it seems to be of significant size and quality. Discussions are already underway on the promotion of Cali's medical services overseas, by attracting patients from Central America and the Caribbean, with particular reference to four areas of excellence—heart, eye, and cosmetic surgery, and kidney transplants. Plans include links to tourist services (providing package tours for "medical tourism"), bank credit, air services, and recuperative hotel services. However, more attention needs to be devoted to gaining internationally acceptable accreditation (through, for example, US health insurance companies), and to the broader question of ensuring adequate security and quality of life in the city.

The local CDS team for economic reactivation has undertaken a survey of higher education as an economic sector. They enumerated 71 institutions of higher education, including 9 universities. The direct and indirect employment from this sector (excluding wider multiplier effects) was estimated at 60,000, or 7 percent of the official labor force. The universities have grown swiftly during the recession—from 20,000 students in 1992 to an estimated 76,000 in 1999. Student and university spending have thus provided a significant counter-cyclical force offsetting the slump. There is still little evidence on the precise ways (and magnitudes) in which this spending feeds through to the rest of the economy—in, for example, consumer spending, transport, construction, electronic equipment, books and stationery, etc. However, the team made a preliminary estimate that the value of gross exports from the sector is US$11 million in 1999 (almost certainly an underestimate, since the flows are so difficult to track). The universities are concentrated in the southern part of the city, where the major sports facilities are also located. This is regrettably distant from the downtown area—close proximity would mean not only that university and student spending would strengthen the city core, but there could be economies of scale in the provision of common facilities, in cultural activities, restaurants, bookshops, etc. However, the emergence of a university quarter in the south could maximize exploitation of the facilities in the area as a single complex, to the benefit of both resident non-university people and outsiders (in terms of culture, sports, non-formal educational facilities such as public lectures, courses, conferences, etc).

In terms of the cultural economy, the Cali team concerned with this topic has begun to discover something of the magnitudes involved. The city claimed (in 1998) to possess: 33 cinemas, 16 theaters, 8 museums, 60 libraries, 48 exhibition centers, and 138 convention halls and auditoriums (figures were not available on radio and television studios and stations, music studios, etc). This provides a substantial hardware component to a cultural economy, generating significant revenues for the city, despite security concerns. The facilities are concentrated in the northern downtown area (10 percent of cinemas, 76 percent of theaters, all museums) and the south (21 percent of cinemas), and these two points of focus provide a physical framework to enhance the general provision of culture (by linking these facilities to the distribution of hotels and restaurants, and to tourism policy). There are important festivals that make use of these and other facilities—for example, the Cali Fair (December 25-31) attracts some 1.5 million participants (with direct and indirect employment, excluding multiplier effects, of some 20,000); and the biennial two-week International Festival of Arts, which in 1999 attracted 12 theater groups (three from abroad) and more than 70 music and dance groups. This suggests something of the vigorous cultural life of the city, as do the numerous cafes and clubs, although there are few reliable figures on the overall economic impact of these activities and facilities, and thus little information on how public, private, or partnership investments could enhance theses effects.

Cali is well-supplied with sports facilities, particularly since the major upgrading undertaken at the time of the Pan American Games in Cali in 1971 (also the occasion of upgrading the airport). Teams participate in 37 sports leagues, which, along with other activities, draw an audience of up to an estimated 19 million per year, and generat an estimated US$10 billion in gross revenues. The football stadium has a capacity of 40,000, and the city provides facilities for the South American volleyball championships, for the Davis Cup, and for world contests in swimming, cycling, and hockey. The largest concentration of activity is in the south (*comuna* 17, with some extension to 18), although additional facilities are in the north (*comunas* 4, 6, 7, 11). As with the cultural economy, sports activities are an additional element supporting the tourism economy, with the same kinds of implications for city accommodations (hotels, etc.), restaurants, retail trade, transport, etc. There are, however, no data on the precise impact of the sports economy on the city's economy at large.

In terms of tourism, the city claims to have 70 hotels and hostels (with more than 6,000 beds), 93 percent of them in *comunas* 2, 3, and 4 (58 percent in *comuna* 3 alone). A future high-growth tourism industry would rapidly exhaust this stock, but it is a starting point. Tourism, though already significant, will not realize its full potential unless security could be guaranteed, and unless the city's facilities are upgraded, the city center and colonial and independence period architecture are restored, etc. But the potential is considerable.

The beginnings of an information industry are already apparent, with both independent software providers and industries such as international publishing. For example, Open Systems provides the bulk of software for Colombia's telephone systems and is expanding abroad (to Brazil). Companies such as these are small employers and highly specialized, but they provide the possibility of a much wider range of tradeable services, especially in the labor-intensive sectors of data loading and processing. There are few signs of these activities at present. However, it appears that the supply of people qualified for such work is expanding rapidly, due to training courses at universities.

The service economy already produces the major part of the output of the city, yet it is striking how little of this sector is known or quantified, particularly in the area of service exports. Yet it is *the* city economy in important respects, and therefore any strategy for city development must begin with this sector. However, the service economy has a peculiar characteristic—it depends economically on the quality of life in the city. It is impossible to offer an export-oriented health sector if visitors become ill from polluted water, or a tourism industry if there are problems of personal safety on the street or during visits outside the city (particularly on the Pacific coast and in the mountains). The longstanding municipal agenda of achieving an efficient system of public services is now a necessary condition for the effectiveness of the new urban economy, which is, in turn, a key component in achieving high employment.

At various points, this account has touched upon important geographical concentrations of activity that merit particular attention in both economic and physical planning: (a) the financial quarter or central business district; (b) the downtown area, with the historic center, cinemas, theaters, antique shops, restaurants, hotels, etc. (this area could be made into a pedestrian mall); (c) the southern university quarter, with its concentration of cultural facilities and sports facilities; and (d) the area with the wholesale markets, linked to warehousing, transport junctions, banking facilities, etc.

The <u>construction sector</u> represented more than 10 percent of Cali's GDP in 1996, and is said to have played an important role in the onset and severity of Cali's recession. Between 1989 and 1994, there was a boom in construction, with US$160 million of new investment,[20] 60 percent of which was in housing.[21] The boom is said to have been partly fueled by the laundering of narcotic funds, which were allegedly used to purchase and develop land (into shopping malls, hotels, restaurants, and other tourism facilities). Whether this is true and on what scale it occurred is probably now of less importance than the economic impact of the boom on the subsequent recession. From a peak employment in the industry in 1994 of 8 percent of the labor force, the industry contracted to a low of 5 percent in 1998. This contraction, as we have seen, made an important contribution to the city's overall unemployed workforce. Another indicator of the crisis in this sector is that the number of building permits issued fell from 1,659 in 1991 to a low 310 in 1997. More recent data from the Construction Chamber indicates that from 1997 to 1999, the number of square meters approved for construction fell by 29 percent. The fluctuation in construction could well have been a force precipitating the downturn and exaggerating the depth of the trough. However, in the longer term, the cyclical downturn will not affect the city's economy as a whole, and the industry is likely to revive with an increase in demand.

The so-called <u>informal economy</u> is difficult to quantify. Except for employment information included in the national household surveys carried by the National Statistical Department (DANE), very little is known about this important segment of the economy. According to DANE's methodology, informal employment is made up by four categories: (a) individuals who are self-employed; (b) enterprises of no more than ten employees; (c) domestic workers; and (d) unpaid family workers. Using this definition, the degree of informality in Cali's employment in the period 1986-1998 is presented below (please refer to Graph 3.4).

Informal employment in the city has been consistently higher than 50 percent over the last 10 years. Up to 1994, when Cali's economy was growing, the level of informal employment decreased steadily, reaching a low of 52.3 percent. With the crisis, informal employment increased sharply, reaching 59.3 percent in 1998.

The degree of informal employment decreases with education (please refer to Table 3.6). Persons with less than ten years of education stand a very high chance (70 percent) of working informally, earning lower incomes than what they could earn in the formal sector. The most educated, in particular, earn higher incomes in the formal sector.

---

[20] Urrea and Mejia, no date.
[21] *El Tiempo, 1996.*

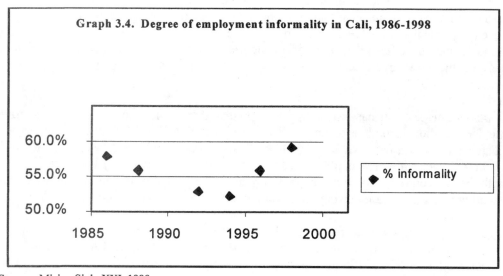

Source: Mision Siglo XXI, 1999.

Table 3.6. Employment informality and education level in Cali, 1998

Years of education	No. of formal workers	No. of informal workers	Informality (%)	Average salary, formal	Average salary, informal	Salary ratio F/I
0-5	45,992	195,727	81	243,069	157,757	1.54
6-10	64,436	144,796	69	284,466	212,801	1.33
11-15	145,957	122,261	45.6	425,433	312,070	1.36
16 +	78,760	25,852	24.7	1,054,254	577,535	1.82
Total	335,145	488,636	59.3			

Source: Mision Siglo XXI, 1999.

Finally, in terms of economic activities, the highest levels of informality are seen in commerce, construction, agriculture, and non-financial services, all of which have informal employment levels above 59 percent (please refer to Table 3.7.). Even the lowest level, 34.1 percent in financial services, is very high. In all cases, *reported* incomes earned through informal employment are substantially lower than those offered in formal jobs.

Exports

The information available on the export sector has the following limitations: (a) there is information only on exports to the rest of the world, not to the rest of Colombia; (b) the recorded exports are from the Valle del Cauca Department rather than from Cali; and (c) the exports concerned are goods, not tradeable services, which is a serious deficiency, since the city is becoming more focused on services. In sum, the available data cover, at most, the exported share of about a third of the city's economy. [22]

---

[22] Colombia has a low dependence on exports (about 13 percent of gross domestic product), compared, say, to Mexico's 35 percent, let alone the much higher ratios of East and Southeast Asia. Thus, the dynamic growth of diversified external markets is not brought to bear to offset domestic recession.

Table 3.7. Employment informality by economic activity in Cali, 1998

Activity	No. of formal workers	No. of informal workers	Informality (%)	Average salary, formal	Average salary, informal	Salary ratio, F/I
Agriculture	3,115	4,943	61.3	581,778	360,861	1.6
Manufacturing.	82,616	84,693	50.6	412,682	227,227	1.8
Construction	11,468	26,046	69.4	426,406	236,715	1.8
Commerce	49,168	162,080	76.7	309,969	228,300	1.4
Financial services	55,757	28,869	34.1	681,866	486,344	1.4
Other services	103,536	152,134	59.5	603,382	174,464	3.5
Total	337,014	491,843	59.3			

Source: Mision Siglo XXI, 1999.

Table 3.8 lays out the changing composition of a selection of the main commodity exports of the Valle del Cauca Department between 1992 and 1996 (unfortunately, taking single years makes it difficult to distinguish trends from short-term fluctuations). The table has been divided into traditional and non-traditional manufacturing, since this gives a rough idea of the relative labor and skill intensities of the output and thus the different employment potentials. The overall cumulative growth during these years, 48 percent, provides the benchmark separating subsectors in relative decline (below this level) or expansion (above this level).

The first observation is that the department depends on manufactured exports that consisting mainly of partly processed raw materials, with relatively low levels of value added. About 40 percent of exports are raw material-based, and are often subject to considerable external price fluctuations. For example, Colombia's coffee exports fell in volume by 16 percent in 1998, as the result of a 30 percent decline in prices. In what we have called traditional manufacturing, sugar is the most important single export (about a quarter of the total), and is expanding moderately (17 points above the average). Very rapid expansion is took place in sweets (241 percent cumulative growth in this period), garments (193 percent), shoes (138 percent), paper and packaging (103 percent), and synthetic textiles (96 percent). At the same time, other foodstuffs and fish were in relative decline, and cotton goods, prepared leather, other manufactured leather, and publishing declined absolutely. Too small to be included here are some other absolute declines—cotton (-91 percent), processed fruit (-86 percent), coffee products (-76 percent), fruit (-46 percent), shellfish (-39 percent), and cotton textiles (-26 percent). Other farm products expanded rapidly (261 percent), with asparagus perhaps a key constituent. Also too small to be included are cut flowers, with very rapid growth (480 percent); and tobacco, which was recorded in 1992 but had disappeared by 1996. In sum, if this period represents the trend, it shows a simplification in traditional manufacture—to sugar, garments, shoes, and a handful of other smaller items. With more accurate data, we might be able to separate out Cali's contribution and the specialization of other particular districts in the Department.

Table 3.8. The changing structure of manufactured exports, Valle del Cauca Department, December 1992 – December 1996

Traditional manufacturing	% share of exports in 1996	Rate of growth (%) 1992-96
Foodstuffs		
Sugar	21.2	+65
Sweets	5.6	+241
Fish	3.8	+38
Other	4.4	+37
Subtotal	37.2	+45
Textiles		
Cotton goods	4.5	-18
Garments	2.6	+193
Synthetic textiles	2.2	+96
Subtotal	10.1	-4
Leather products		
Shoes	1.7	+138
Prepared leather	0.5	-29
Other leather products	-	-35
Subtotal	2.8	+26
Paper industry		
Wood products	0.5	+140
Paper and packaging	9.9	+103
Publishing	4.2	-35
Subtotal	14.0	+23
Non-traditional manufacturing		
Metals		
Cast iron	5.9	+85
Electrical equipment	1.9	+65
Machinery	1.6	+91
Metallurgy	0.8	+43
Subtotal	10.8	+71
Chemicals and others		
Chemicals	5.9	+92
Pharmaceuticals	3.4	+456
Insecticides, disinfectants	0.6	+327
Subtotal	9.9	+160
Plastic and others		
Synthetic materials	5.6	+66
Rubber	4.6	+198
Plastics	0.2	+1,082

Petrol and derivatives	-	3,106
Subtotal	11.6	+107
Minerals		
Non-metallic mineral products	0.06	+20
Coal	-	+299
Subtotal	0.3	+63
Other industrial products	2.6	+55
Total	100.0	+48

Source: Calculated from *Plan Regional de Exportationes del Valle de Cauca: Frente a la Comunidad Andina de Naciones,* in Secretaría de Fomento Económico y Competitividad, 1998 (b).

The picture of exports of non-traditional manufactures is generally expansionary, suggesting the that the department is moving toward exports with higher value added and more skill-intensive output. The highest rates of growth have occurred in chemicals and synthetic materials—plastics (1,082 percent increase), pharmaceuticals (456 percent), insecticide and disinfectants (327 percent), and rubber (198 percent). Coal exports also have expanded rapidly (299 percent). In the metals fields (including electrical and mechanical machinery), the growth has been respectable, although metallurgy is in relative decline. Non-metallic minerals—with the exception of coal—are also in relative decline. Transport equipment (0.5 percent of exports in 1992) shrank by half during the period, but hand tools—too small to be included here for either year, grew by a spectacular 12,000 percent. Thus, in sum, the department's particular strength appears to be in chemicals, followed by some selected items of metal manufacture.

These figures cover only a fragment of the department's exports, and before the full force of the current recession strained the economy. It is recession which, other things being equal, eliminates the weaker sectors of the economy and rewards with survival those with relative strengths. Nonetheless, the breakdown gives a suggestion of the emerging advantages in export manufacture.

There are unrecorded exports on which there is little information. With a land border with Ecuador within reachable distance (Venezuela and Ecuador take about half Colombia's manufactured exports), it is possible that the unrecorded exports affect the department and the city. There are also unrecorded exports of illegal goods, especially narcotics, which may also affect transactions in the department. In the past, some have suggested the inflow of narcotics earnings to Cali may have had a Dutch disease effect, expanding the domestic market and thereby discouraging manufactured exports. (On the other hand, drug revenues are said to have financed smuggled imports, affecting the markets for domestic manufactures.) The destruction of the Cali cartel may have affected this phenomenon, thus exaggerating the scale of the recession; or it may have merely dispersed drug revenues to other locations.

Infrastructure

Social infrastructure, particularly education, is discussed in Chapter 6. This section discusses physical infrastructure, focusing on power and transport, which are crucial to the city's capacity to earn income. In the case of power, plans are underway to relieve existing bottlenecks by opening new hydroelectric (Calima III) and thermal power facilities. In the case of transport, the problems may be more severe.

Colombia's land transport system is well known for being high cost. Compared to much of the rest of Latin America, there is a relatively low number of kilometers of paved roads per million population, and a low penetration of mountainous areas compared to countries with similar terrain. Partly as a result of this, transport costs are between 25 and 75 percent higher than comparable mountain countries in the region. Administrative bottlenecks exacerbate this problem, so transport costs add some 30 percent, on average, to the cost of exported goods (compared to 10 to 20 percent in Europe and North America), and thus tend to eliminate any comparative advantages Colombia might possess. In the past, poor transport might have worked to give Cali a measure of protection compared to other centers of the country, but as we see in the next section, this is no longer the case. Colombia's Caribbean ports now compete with Buenaventura on the Pacific coast for cargo from the Valle del Cauca Department. Around a million tons of cargo are also carried by road to Ecuador annually from various areas in the department. In general, therefore, road and rail transport are important constraints on Cali's capacity to compete with its nearest rivals.

The Buenaventura port was for a long period a major bottleneck for exporters and importers in Cali and the rest of the Valle del Cauca Department. In the 1950s, the port had been the largest in Colombia, handling 60 percent of national exports by volume in 1955-60, and half of national imports in 1960. This was reduced to 20 percent by 1990, the result of poor investment in unfavorable physical conditions—as the average size of cargo ships increased, they became too large for the main channel. In addition, in the past, poor management of the port raised costs and increased delays, pilferage and breakage. Since then, privatization of Colombia's ports has led to increased investment and rationalization—the direct and indirect labor force of 10,000 in 1990 was reduced to just over 4,000 in 1996.[23] Efficiency was sharply increased, but costs remained high—the 1995 Monitor Report on Cali quoted (without sources) costs per ton of cargo bound for New York as US$40 via Buenaventura and $5 via Cartagena, implying that port handling charges accounted for a large part of this difference.

Nonetheless, the port has reclaimed an increasing proportion of national cargo (60 percent of the throughput consists of sugar cane and wood from the Valle del Cauca Department), even if there some cargo is still drifting northward to the safer and quicker ports of Barranquilla and Cartagena—a trend that could be further enhanced by the development of all-year navigation on the Magdalena river and, possibly, the opening of a new tunnel on the Bogotá-Cali highway. At the same time. Buenaventura is seeking to deepen and maintain the main channel (to 30 meters depth) to accomodate larger vessels, and to develop an industrial zone and a new deep water container terminal on the Aquadulce peninsula. These changes may go some way toward restoring the port's preeminence, although the context has now been transformed by privatization and the emergence of nine new private ports to compete with the five established ports, as well as by the competition of other Pacific ports (in Chile, Peru, and Ecuador). However, the main problem for the port of Buenaventura is perhaps not the dock facilities themselves, but the road access through the mountains, particularly that part of the road where both Bogotá and Cali cargo are carried. A rail line would ease this so far as Cali is concerned, but the proposed rehabilitation of the rail connection between the port and Cali appears to be trapped in litigation over the use of the land on either side of the line. In any case, the highway will ultimately have to be improved if the full potential of the port is to be realized.

The development of cheap air freight could go a long way toward offsetting the disadvantages—including insecurity—of the Buenaventura route. It could be the basis for new high- value/low-weight manufactured exports. In 1998, the airport handled two million passengers (13 percent of them international) and 66,000 tons of cargo (65 percent international), with international connections to Miami, Panama, and Quito. The costs, however, are high—for example, rates for exported publications recently quoted as US$0.08 per kilogram by sea and US$2.50 by air. There are also problems of security (increasing insurance costs, delays, and the risk of seizure of aircraft)—for example, in August 1999, narcotics smugglers were arrested as they tried to use an American Airlines flight to move half a ton of heroin to Miami. The imminent privatization of the airport may begin to improve its efficiency and lower freight rates, thus

---

[23] In 1992, *El Pais* claimed that the unemployment rate in Buenaventura was 65 percent. (22nd March)

allowing Cali's exporters to escape some of the problems of existing road, rail and sea exits, as well as stimulating new exports of goods and services.

Given all of these circumstances, it would be useful to explore ways to rationalize all of Cali's transport facilities, which are so crucial for the city's economic future. A dry dock—with customs services and the nonstop through-shipment of in-bond cargo to airports or seaports—could be located near the airport, close to the proposed rail links and to road junctions, thus minimizing the costs and the delays in transhipping cargo between transport modes. The concentration of activity could also be exploited for a business park of firms associated with the movement of cargo.

Proposed strategic priorities for Cali's economic recovery

The city's economy is in the process of structural change, and while we can see the dispersal of manufacturing away from the city, some obvious strengths have been noted—the dynamic growth of exports in garments, shoes, cut flowers and asparagus, and printing and publishing. The city's software industry, while very small still, also has much potential. The severe disadvantages have been noted, particularly those which cumulatively make impossible a significant tourism industry. The distinctive advantage of the city is its geographic location, commanding the southwest corner of the country between the Andes and the Pacific, and with access to the Pacific-American economic region. The coast also has created a unique culture, particularly in music, which could provide a basis for tourism in the city. Around that core, reproducible tradeable services could be grouped to generate a strong employment basis, particularly if facilitated by a relatively educated labor force. All this, however, requires the establishment of security, an upgrading of social and physical infrastructure, the rehabilitation of parts of the city of particular interest to visitors, and an improvement in the quality of life in the city. The following proposals for achieving these goals are very preliminary, and will undoubtedly benefit from the continuing work of the local counterpart team.

- *Generate employment through labor intensive industries.* Two courses of action are recommended. The first is to develop, with the Chamber of Commerce and relevant business associations, export strategies for Cali's labor-intensive industries, namely shoes, garments, food processing, and paper and packaging. The second and complementary action is to expand efforts to develop training facilities, especially for young workers (especially young women) who already have a measure of education, and for migrants (especially those located in squatter areas where training facilities already exist). The municipality has a training program of this sort, but because of budget limitations, it has partial coverage. The training must be related to identified sectors of labor-intensive manufacture and directed at exporting. With the city's continued efforts of the city to characterize economic activity by *comuna*, it will become possible to tailor such training to particular city localities. The combination of these two measures should generate employment opportunities in the medium term, particularly for the young.

- *Develop a portfolio of select tradeable services.* The city has some clear strengths in terms of services. What needs to be done at this time is to select a few of these and concentrate on improving their output.

First, in terms of finance, the planning authorities should, in a more determined way, seek to encourage the development of a financial quarter as a key component of a central business district, which may or may not coincide with the downtown area. Many plans exist, but so far there has not been a true partnership with the private sector and the Chamber of Commerce to take this idea one step further.

Second, for the health sector, the Chamber of Commerce and national export agencies are already developing an export strategy for the four areas of excellence in Cali's medical services—and this will include issues of marketing, bank credit, transport, and possible accreditation (for example, by securing the acceptance of US insurance companies). The city needs to identify more clearly where these medical clusters are likely to be; to what extent those areas need a physical upgrading program (with redesign of the urban fabric, the addition of squares, parks, vegetation); and how ancillary services (convalescent homes, cultural facilities, pharmaceutical services) can be developed.

Third, in terms of higher education, Cali's excellence in several areas of teaching and research can be used to attract significant numbers of students from outside the city, with strong multiplier effects on the city's economy and labor force. The city's universities, colleges, and other institutions of higher education could collaborate in this effort. In addition, many of these institutions have facilities that could strengthen the city's provision of culture (theaters, concert halls, lecture halls etc) and sports (stadiums, swimming pools, race tracks). This complex of facilities, much of it located in the southern university quarter (*communa* 17 and part of 18), could be complemented by restaurants, small hotels, galleries, museums, etc., to attract tourists and out-of-town visitors.

Fourth, culture and sports are frequently seen as elements of local consumption, although their capacity to attract outsiders is a key component in the market for city goods and services, particularly in the university quarter and the old town center in the north. The city center is slowly recovering from long years of neglect, and is a strong comparative advantage for the city, provided that the architectural and cultural heritage of the area is fully restored and security can be guaranteed. Pedestrianizing the area would help the process of redevelopment and increase its attractiveness to visitors; excluding motorized traffic would allow the streets to be recovered and landscaped, and would make possible the free movement of street traders, musicians, etc. This would, in turn, attract many more shops, restaurants, museums, galleries, etc.

Finally, the city needs to develop a skilled labor force to support the growth of the information industries, particularly in labor-intensive sectors such as data loading and processing. Along with the growing number of university courses, this will entail low-level introductory and advanced training courses in these areas, especially in squatter settlements. As the Indian case shows, a skilled labor force is the key to the growth of these industries.

Some of these recommendations coincide with the proposals in Chapter 6, Education.

- *Improve the city's infrastructure.* A program for economic expansion of the city presupposes intensified efforts (provision of adequate water supplies, solid waste disposal, efficient transport services, adequate security) to create an acceptable quality of life. In particular, the transport network connecting the city to external domestic and foreign markets is crucial to the city's prosperity. Upgrading the transport network will involve: (a) completing the Calima III hydroelectric and other projects to ensure adequate energy supplies; (b) completing the rail link to Buenaventura port; (c) upgrading the road link to Buenaventura, and in particular, expanding facilities from the point where freight movements from Cali and Bogotá coincide; and (d) upgrading the airport and reducing air freight charges. Ensuring adequate security of cargo handling at the airport provides a transport link that is less vulnerable to interruption than road and rail links. With such security, the following investments would make sense: a dry dock (so that in-bond cargo can move either by air or via the seaports), an intermodal junction point, a business park, and/or an export processing zone or free trade zone.

- *Invest in information.* Finally, if city stakeholders are to play a continuing role in guiding the city, information becomes crucial. The capacity to monitor the city economy is currently handled by the

Chamber of Commerce, and given the available official data, this is done quite well. But much more needs to be known about the city's export economy, informal economy, and tradeable services economy. To this end, a small think tank could be set up to provide the municipal authority, the Chamber of Commerce, and the universities with research and studies.

At the time this document was being revised for publication, several institutions in Cali came together to create an Economic Observatory for the Valle del Cauca Department, which plans to produce quarterly reports. The first report, published in 2000, crossing sector outputs, employment data, geographic location, and time trends, and contains a great deal of information on the informal economy.

4. SOCIAL DEVELOPMENT

Introduction

Poverty in Cali increased from 29.8 to 39.0 percent between 1994 and 1998, and the percentage of those living in misery—defined as not being able to purchase a basic basket of food—nearly doubled, from 5.3 to 10 percent. The better-off in the city were able to protect their well-being and maintain their standard of living; consequently, income inequality increased as well (Cali is the only large city in Colombia where this occurred). With rising unemployment, poverty, and inequality, the lives of the population in marginalized areas are becoming bleaker: child labor and secondary school dropout rates are on the rise, as is the threat of violence from various youth gangs. At the same time, shrinking tax revenues have strained the city's budget and comprised the functioning of basic social services.

Although the social situation is bleak, it is not hopeless. Cali has a tradition of civic engagement, public-private partnerships, and a lively academic community heavily engaged in city affairs. Moreover, groups from various areas and income quintiles all agree on priorities for the city's rehabilitation, which makes possible the formulation of an integrative social policy.

This chapter is based on a number of background studies[24] and a rapid city household survey of service access and satisfaction,[25] which proved a rich source of information on social conditions in the city. The chapter is organized as follows: The next section provides background information on the development of poverty and inequality in Cali over the past few years, comparing it to developments in other large Colombian cities. The third section draws largely on the household survey to present a poverty profile for the city in 1999. Section four reports on the level of major municipal services (education, health, other social programs), while section five reports on how the population of Cali rates and prioritizes services. Section six includes several broad suggestions for the formulation of social policy.

Cali in the national urban perspective, 1994-1998

Social indicators show how much the city recession has impacted the lives of the population (please refer to Table 4.1). Unemployment developments follow economic dynamics in the city very closely. Traditionally, unemployment has tended to be above the national average, due to a higher demand for skilled as opposed to unskilled labor. During the national economic downturn from 1994 to 1998, open unemployment in Cali rose by 8.4 points and now affects almost one out of every five Caleños in the labor force. This rate is far higher than in other large cities in Colombia. Unemployment has many repercussions—poverty, most directly, but also rising social tensions and violence. Today, a quarter of a million people are actively looking for a job in Cali.

Cali is the only large city in Colombia to show an increasing concentration of income, and an increase in the Gini coefficient of inequality, between 1994 and 1998 (please refer to Table 4.1). Other cities, notably Barranquilla—although equally hit by the economic downturn—showed improvement in the distribution of income in this time period.

The increase in unemployment has been highest for the poorer income groups, and rose most strongly for the bottom two income quintiles (please refer to Table 4.2). This resulted in several negative effects. First, the labor market has become increasingly informal; it is estimated that more than half of all jobs are now of an informal nature. While the informal sector provides an important alternative to unemployment,

---

[24] Jerozolimski, 1999; Santamaria, 1999; Urrea and Ortiz, 1999.
[25] *Encuesta De Acceso y Percepción de los Servicios Ofrecidos por el Municipio de Cali* (EPSOC), 1999

it does carry a weight as the tax base of the city shrinks and workers' access to social security diminishes. On average, access to social security, health care, and retirement funds declined in Cali during the period (please refer to Table 4.3), and all three developments were worse than the national urban trend. Second, rising unemployment resulted in a marked increase in child labor over the last few years; again, Cali is the worst of the cities considered here (Table 4.3).

Table 4.1. Cali in the national perspective

City	Unemployment			Inequality		
	1994	1998	Change	1994	1998	Change
Bogotá	4.9	14.8	+9.9	0.55	0.54	-0.01
Medellin	8.6	16.7	+8.1	0.58	0.51	-0.07
Cali	11.3	19.7	+8.4	0.51	0.54	+0.03
Barranquilla	10.1	13	+2.9	0.76	0.57	-0.19
Bucaramanga	8	15.9	+7.9	0.48	0.48	0
Total urban	7.4	15.7	+8.3	0.58	0.55	-0.03

Inequality defined by the Gini coefficient.
Source: Santamaria, 1999.

Table 4.2. Unemployment in Cali, 1994 and 1998

Indicators	Income quintile					
	1	2	3	4	5	Total
Unemployment rate, 1994	24.8	14.3	12.5	10.4	5.5	11.3
Unemployment rate, 1998	43.5	26.2	15.9	13.2	6.9	19.6

Source: Santamaria, 1999.

Table 4.3. Social security access in major Colombian cities

	Health care			Retirement funds			Child labor (12 to 16 years)		
	1996	1998	Change	1996	1998	Change	1994	1998	Change
Bogotá	56.2	55.5	-0.7	40.1	42.7	+2.6	8.1	7.7	-0.4
Medellin	64.7	62.6	-2.1	44.0	49.9	+5.9	7.5	6.6	-0.9
Cali	52.6	49.2	-3.4	39.5	37.8	-1.7	7.9	9.1	+1.2
B/quilla	44.3	38.6	-5.7	31.3	30.8	-0.5	6.1	5.9	-0.2
B/manga	44.6	48.9	+4.3	33.2	37.2	+4.0	15.8	14.9	-0.9
Total Urban	55.5	52.4	-3.1	39.7	40.3	+0.6	8.2	8.4	+0.2

Source: Santamaria, 1999.

Closely linked to increasing unemployment, both poverty and misery increased strongly over the past four years. A careful evaluation of the national household surveys from June 1994 and 1998[26] finds (as noted above) that poverty in the city, using the poverty line defined by the Colombian Statistical Institute (DANE), increased from 29.8 percent to 39.0 percent, while the percentage of the population in misery rose from 5.3 percent to 10 percent. At the same time, natural population growth and migration added another 160,000 citizens to Cali, bringing it above the two million mark (2,070,000 inhabitants in 1998). In total numbers, this means that more than 800,000 people were living in poverty in the city in 1998, and more than 200,000 were living in misery.

---

[26] Urrea and Ortiz, 1999.

Table 4.4. Poverty and misery in Cali, 1994 and 1998

Indicator		1994	1998
Poverty	Headcount rate (%)	29.8	39.0
	Absolute number ('000)	571	807
Misery	Headcount rate (%)	5.3	10.0
	Absolute number ('000)	101	207

Source: Urrea and Ortiz, 1999.

Poverty in Cali

We used household per capita income as our welfare indicator for the purpose of this study.[27] The labor market and income module of the household survey is relatively limited, but it does distinguish among a number of different income items: wages and salaries of primary and secondary occupation (by unit and frequency of payment), additional income from work, interest income, rental income, pensions, transfers, and other sources. We found, however, that the data base did not include wage and salary income for about one third of informants who worked during the time period of the survey. For these, we constructed a simple estimation model of hourly wages as a function of individual (education, experience, gender), occupational, and area characteristics to impute the missing data. Aggregating all family incomes and dividing by the number of members in the household provided us with household per capita income, which we used to construct income quintiles for the population of Cali.[28] This section covers three main aspects of poverty in Cali: (a) the relationship between income and socioeconomic strata; (b) a spatial analysis of income and poverty; and (c) specific characteristics of poverty in Cali.

In Cali, as in Colombia in general, much of the geographic discussion of poverty is framed in terms of socioeconomic strata. These are socioeconomic groups identified largely from census data, including housing characteristics (material of the house, access to services), which provides information about unsatisfied needs. Strata are defined at the *manzana* level, which generally includes one complete block of houses (four sides); thus, differences in housing and basic service characteristics within a *manzana* cannot be identified. Strata definitions in Colombia play an important role in targeting subsidy programs and setting rates for public services.

---

[27] The EPSOC survey was designed not to measure household income accurately, but rather to obtain a good approximation to facilitate the ranking of households in the city from poor to rich. Therefore, we have abstained from using the survey to derive absolute levels of poverty , based on nutritional measures. Rather, we have used the most recent estimate of poverty from Urrea and Ortiz (1999) to define poverty as the bottom two quintiles, which encompass close to 40 percent of the population. Similarly, we have defined extreme poverty as constituting the bottom 20 percent. EPSOC was also designed to achieve representativeness for five different areas in the city. However, we present estimates at the community level, as poverty levels within the larger areas appear to vary significantly. While we lose precision in the estimate (standard errors increase), we gain a more complete picture of poverty. Standard errors of the headcount poverty rates are included in (Table 4.6). As can be seen, the standard errors are within a reasonable range, except for *comunas* 3, 4, and 12. Thus care is required in drawing conclusions from a rank ordering of these *comunas*. In dividing total household income by household size, we assumed that no economies of scale in consumption exist. Recently, this assumption has been questioned (see Lanjouw and Ravallion, 1995).

[28] In dividing total household income by household size, we assumed that no economies of scale in consumption exist. Recently, this assumption has been questioned (see Lanjouw and Ravallion, 1995).

> **Box 4.1. EPSOC: Encuesta de acceso y percepción de los servicios ofrecidos por el Municipio de Cali, 1999.**
>
> With financial support from the Bank-Netherlands Partnership program, the *Encuesta De Acceso y Percepción de los Servicios Ofrecidos por el Municipio de Cali* (EPSOC) was conducted in Cali in September 1999. The survey built on a city module being piloted in Kampala, Uganda, and was adopted to the situation in Cali with the help of the *Universidad del Valle, Centro de Investigaciones y Documentación Socioeconomicas*. Data collection and tabulation were carried out by an experienced Colombian survey firm, *Centro Nacional de Consultoria*. The survey covered 1,912 households in the city that were representative of five different areas and six socioeconomic strata, as defined by the Colombian statistical institute (DANE). The value added of the survey (compared to existing data sets) was that it was tailored to the city and combined quantitative information (such as household income) with qualitative information on the population's priorities and satisfaction with programs. The survey had modules on housing; access to and satisfaction with basic services, education, and health; the labor market, food security, participation in city affairs, and priorities of the population.
>
> *Source*: Hentschel, Mehra, and Seshagiri, 1999.

We are interested in looking at the relationship between socioeconomic strata, as defined by the statistical institute, and income per capita as an alternative welfare measure. Differences might arise between these measures for two reasons. First, within a given *manzana*, considerable heterogeneity might exist among different households, so that the "average" stratum is only an incomplete description of inequality within a given neighborhood. Second, household income per capita and strata might not be very closely related. Especially in an urban setting, the type of building material or access to basic services might not be a close predictor of household income; education and employment status could be closer proxies.

Indeed, income and strata show a relatively high variation. Table 4.5 shows the relationship between the two variables. The strata (row entries) are mapped by income quintiles—with 1 being the poorest 20 percent of the population and 5 being the 20 percent with the most income per capita. The last column entry shows the contribution of each stratum to the total population; for example, 20.4 percent of the population in Cali are mapped as stratum one, while 30.6 percent are stratum 2. We are interested in how income varies across strata. As can be seen, the variability is very high, especially for the population in strata 2 and 3—in stratum 3, income is almost evenly distributed across all quintiles. Since strata are the criteria used to determine tariffs for public services and to target subsidies, it is worrisome to see that, at least in Cali, stratification does not coincide with income distribution. This result cautions against using strata as the only identifying variable for geographic targeting of beneficiaries, and shows the need for a review of the strata identification process. For this reason, we use income figures rather than socioeconomic data to analyze spatial differences in poverty.

Table 4.5. Income and socioeconomic strata, Cali, 1999

Strata	Income quintile					Total	Contribution
	1	2	3	4	5		
One *(bajo-bajo)*	30.4	28.8	20.0	15.5	5.4	100	20.4
Two *(bajo)*	23.8	24.9	24.1	16.4	10.9	100	30.6
Three *(medio-bajo)*	14.5	15.9	22.1	25.7	21.8	100	33.5
Four *(medio)*	13.4	8.7	9.7	23.6	44.6	100	10.2
Five *(medio-alto)*	5.6	5.3	3.5	14.5	71.2	100	4.8
Six *(alto)*	5.6	3.5	7.1	18.5	65.2	100	0.5

Note: Totals may differ from 100 percent due to rounding of individual figures.
Source: EPSOC, 1999.

Colombia is a highly heterogeneous nation, and so is Cali. As central areas modernize and increase in value, the lower-income population settle in marginal areas with precarious environmental and infrastructural conditions, and with insufficient or substandard services, transportation, and urban amenities. In those areas, multiple deprivations are concentrated (income poverty, low education, health risks, insecurity, etc.).[29] The poorest area in the city, *Aguablanca,* comprises *comunas* 13, 14, 15, and 16 (although the latter is sometimes not counted as a part of Aguablanca by its residents) at the eastern edge of the city (please refer to Map 4.1).

Table 4.6. Poverty headcount rates by *comuna*

Comuna	Headcount poverty	Headcount extreme poverty	Contribution to total poverty
1	54.5 (6.71)	14.7 (4.77)	2.8
2	15.3 (2.11)	11.8 (1.89)	1.7
3	42.0 (8.34)	19.5 (6.70)	2.4
4	45.3 (10.61)	14.6 (7.53)	1.5
5	19.0 (4.53)	10.9 (3.60)	2.6
6	51.2 (4.47)	23.4 (3.79)	9.4
7	47.3 (4.99)	31.8 (4.66)	8.8
8	31.2 (4.88)	11.3 (3.34)	4.8
9	22.4 (4.81)	12.2 (3.78)	3.1
10	25.9 (4.28)	12.9 (3.27)	4.1
11	36.2 (4.93)	15.8 (3.74)	7.0
12	28.9 (10.14)	17.3 (8.46)	1.3
13	42.9 (4.52)	27.0 (4.05)	6.4
14	72.8 (3.63)	37.0 (3.94)	15.9
15	55.7 (5.55)	32.7 (5.24)	6.0
16	55.4 (3.99)	24.2 (3.44)	7.7
17	6.7 (3.37)	3.7 (2.55)	0.4
18	53.1 (5.12)	26.4 (4.52)	7.3
19	19.9 (2.95)	7.8 (1.98)	3.4
20	45.7 (6.72)	15.3 (4.85)	3.4
TOTAL	40.0	20.0	

Note: Standard errors in parentheses.
Source: EPSOC, 1999.

As Table 4.6 shows, even within Aguablanca, differences in poverty are obvious: *comuna* 14 is by far the poorest in the city, with a poverty rate of 72.8 percent, while the other two have somewhat lower levels. The *comunas* of Aguablanca have been heavily populated by black migrants from the Pacific coast. The second area commonly identified as very poor is *Ladera*, comprising *comunas* 1, 18, 20 in the western part of the city. These *comunas* are steeply sloped at the hillside and comprise mainly poorer *mestizo* migrants. From the individual poverty rates, we find that *comunas* 1 and 18 have somewhat higher poverty levels than *comuna* 20. The *comunas* in the *northeastern part of the city* (6 and 7), which also include recent migrants, show moderate to high levels of poverty, especially in the *comuna* 6, farthest to the north. A large group of centrally located *comunas* (3, 4, 5, 8, 9, 10, 11, 12) are mostly populated by the lower-middle income class and show moderate to medium levels of poverty throughout, with the exception of *comuna* 5, which has low levels. Lastly, a corridor running from the north (*comuna* 2) through the center (*comuna* 19) to the south (*comuna* 17) is populated by the medium and higher-income classes, and consequently shows relatively low levels of poverty.

---

[29] Jerozolimski, 1999, p. 8.

Although these descriptions make it possible to identify a number of different areas of poverty within the city, heterogeneity within areas and even within *comunas* is high. There are pockets of extreme poverty in specific *comunas* that do not appear to have a high poverty level. Table 4.7 shows a ranking of *comunas* if ordered by the headcount index of poverty (left column) and by extreme poverty (middle column).[30] As can be seen, *comunas* 7 and 13 appear very high in the extreme poverty ranking but are not within the top six according to poverty alone. The existence of such pockets of extreme income poverty within even non-poor *comunas* indicates the need for an extensive information system that allows the municipality to channel resources to these especially poor neighborhoods.

Table 4.7. Rank order of *comunas* by poverty, extreme poverty, and number of poor, 1999

Rank	Poverty rate	Extreme poverty rate	Number of poor
#1	14	14	14
#2	15	15	6
#3	16	7	7
#4	1	13	16
#5	18	18	18
#6	6	16	11

Source: EPSOC, 1999.

While poverty rates by *comuna* are important, of equal importance is the geographic distribution of the poor. Since *comunas* vary widely in size—from a population of 61,000 in *comuna* 4 to 254,000 in *comuna* 14—these two ways of looking at the same picture show different results. Visually, Map 4.2 shows the distribution of the poor population in the city, with the size of the circles indicating the size of the poor population group in each *comuna*. While by far the largest number of poor people lives in *comuna* 14, which has the highest poverty rate (about 185,000 poor people), Table 4.7 shows that *comuna* 6, ranks second in its contribution to poverty.

Two additional aspects are worth noting. First, the *comunas* with the highest headcount poverty rates are also those with the most rapid population growth rates in the last decade—which explains why these *comunas* have the highest share of migrant population.[31] Second, in these rapidly expanding *comunas* the population density is the highest, and crowding and congestion are the consequences.

In terms of <u>poverty characteristics in Cali</u>, six issues are covered in the following sections: high risk groups, food insecurity, housing and access to basic services, school attendance, health care, and violence. The young tend to be most affected by poverty in Cali (please refer to Graph 4.1.). Poorer families tend to be larger and have higher dependency ratios (the ratio of inactive to active household members), so children are at the highest risk of poverty of all age groups. About half the children in the city are not covered by the social security system, and four out of every ten children below seven years do not participate in any program that monitors their development and growth.[32] Adolescents are at a higher risk than average, and those without work show critically high levels of income poverty.

The less educated, on average, also have a higher risk of being poor As Table 4.8 shows, educational attainment is positively linked to income, as expected. It is remarkable, however, that the difference in educational attainment between the first and the fourth income quintiles varies by less than two years. As

---

[30] We define *extreme poverty* here as the population group in the lowest income quintile in Cali.
[31] According to Urrea and Ortiz (1999, p. 7), the population growth rate from 1993 to 1998 was highest in *comunas* 20 (7.2 percent), 18 (7.2 percent), 13 through 16 (2.4 percent), 6 (2.4 percent), and 7 (2.4 percent).
[32] Foro Nacional por Colombia, 1999.

will be discussed in Chapter 6, the pattern of unemployment falls especially heavily on those men and women with incomplete secondary education—more so than on those with a primary education alone.

Map 4.1. Poverty in Cali, 1999 headcount rates

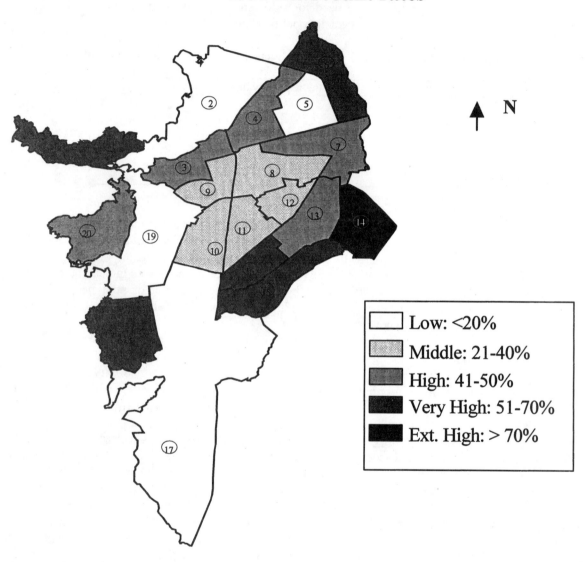

Source: Departamento Adminsitrativo de Planeación Municipal de Cali, Own calculations, based on EPSOC, 1999.

Map 4.2. Distribution of poverty, 1999

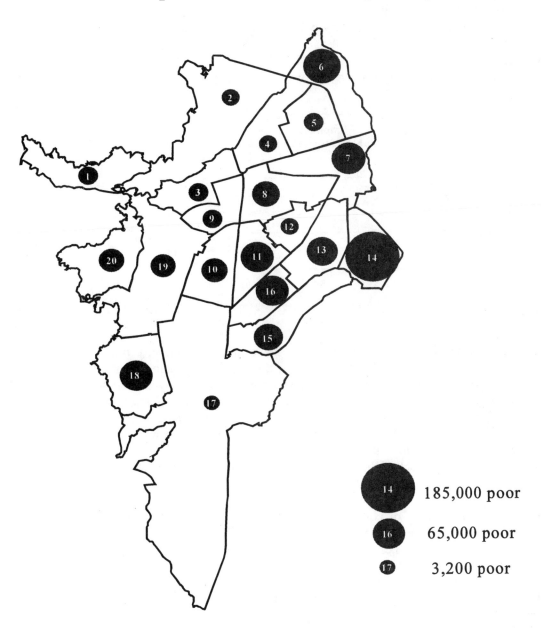

Source: Departamento Administrativo de Planeación Municipal de Cali, Own calculations, based on EPSOC, 1999.

At a somewhat higher risk of poverty is the older population (Graph 4.1), many of whom are without pensions or other social safety nets. Poverty is heavily concentrated among the migrant black and *mestizo* populations.[33] In terms of gender, the likelihood of a female-headed household being poor is about 25 percent higher than the risk for a male-headed household. Finally, the homeless population is also a vulnerable group; many of them recent migrants as well.[34] This population group is not captured by household surveys.

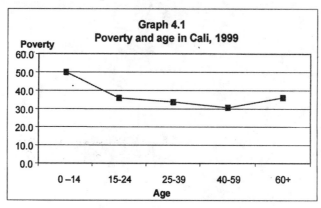

Source: EPSOC, 1999.

Poverty goes hand in hand with food insecurity—and the degree of such insecurity in Cali is at surprisingly high levels. As reported in Table 4.8, about one third of the population in the poorest income quintile had family members who were hungry at least once during the 1998-99 year, and did not have the financial resources to purchase food. And a full 60 percent of such poor parents say they had to reduce nutrition for their children because of insufficient resources over a one-year period. Even though these figures might be somewhat upwardly biased,[35] the results point to a serious problem of food insecurity among the poorest of the poor in Cali. This is especially worrisome since the reach of nutrition programs is extremely small. Only 4.4 percent of the poorest population group report having access to food aid (Table 4.8).

With the geographical concentration of poverty comes differences in quality of housing and access to basic services. In 1996, the Cali Municipal Services Company (EMCALI) found that about 157.000 people lived in substandard housing; i.e., shelters made of wood, adobe, or bamboo.[36] This deficit has been reduced somewhat over the past three years and now affects about 120,000 people. But about 295,000 people still live in houses with unfinished cement floors, and another 20,000 live in shelters with dirt floors. About a third of the population live in rented houses, and ownership is not necessarily linked to material well-being. Close to 300,000 people live in houses without a proper title, 60 percent of whom are poor.

---

[33] EPSOC included questions on racial background, with the surveyor being asked to fill in this question. Though about half the entries for this question were not filled in, previous studies have clearly established the links among race, migration, and poverty.

[34] Urrea and Ortiz, 1999.

[35] The results reported in the table potentially have an upward bias since some respondents in the survey might have thought that answering yes to food insecurity questions would have given the household a chance to access subsidy programs.

[36] Foro Nacional por Colombia, 1999.

Electricity connections and garbage collection are almost universal, but access to water and toilet facilities show marked variation by income group. While almost all of the population have access to public water, almost a third of the poorest income group share a water faucet among several households. Similarly, 320,000 Caleños do not have the use of a private toilet, but share such facilities with neighbors. It is well known that such sanitary conditions come with problems of hygiene and therefore bacterial diseases, especially when the families involved do not have direct and immediate access to clean water.

Table 4.8. Characteristics of income poverty, Cali, 1999

Characteristic	Income quintile					Total
	1	2	3	4	5	
Labor market						
unemployment rate	35.9	22.4	18.4	11.8	5.8	17.1
Education						
yearrs of educ. of household head[1]	6.4	6.6	7.3	8.4	10.3	8.0
Food security						
Family member w/hunger[2]	34.2	22.8	16.9	11.7	5.1	18.1
Had to reduce food intake	60.6	53.6	42.7	25.7	12.9	42.4
Access to nutrition program	4.4	3.7	3.1	1.9	0.5	2.7
Housing						
Rented	37.9	41.0	38.2	36.0	35.5	37.7
Titled	77.5	79.2	89.2	92.8	93.0	86.9
Cement floors, crude	25.4	20.8	13.2	7.1	3.1	13.9
Access to basic services						
Electricity connection	99.5	100	100	100	100	99.9
Hygiene facility	93.9	98.8	98.6	99.6	100	98.2
- in house	73.3	81.1	84.3	89.6	95.1	84.7
Public water	99.7	99.2	99.8	99.8	100	99.7
- single use	69.7	77.9	81.7	86.0	94.9	82.1
- shared use	30.3	22.0	18.3	13.8	5.1	17.9
Garbage collection	95.9	99.0	97.3	99.5	99.8	98.3
Health						
Sick in last 4 months	28.9	24.3	26.6	21.3	19.3	24.1
Used medical facility						
- Public health post	37.6	29.1	27.7	13.6	7.5	22.7
- Public hospital	21.1	16.1	16.9	8.8	8.9	14.2
Health insurance						
- ISS[3]	13.8	19.6	26.3	27.2	29.4	23.3
- EPS[4]	13.2	22.9	20.2	36.4	41.4	26.8
- SISBEN[5]	22.7	12..2	10.9	4.8	0.8	10.3
- not affiliated	46.0	41.2	39.2	26.9	17.6	34.2
Violence						
Family member been victim of assault, robbery, or violent act	20.2	23.8	19.7	20.6	25.4	21.9

[1] Mean education years for the whole population 18 years and older.
[2] in the past year.
[3] ISS = *Instituto de Seguridad Social de Colombia.*
[4] EPS = *Empresa Privada de Salud.*
[5] SISBEN = *Selección de Beneficiarios de Programas Sociales.* SISBEN is not a program per se, but a system that allows identification of beneficiaries for social programs.
Source: Own calculations, based on EPSOC, 1999.

The poor have a higher demand for health care than do better-off income groups. The poorest income quintile has a self-assessed rate of illness of around 30 percent (over a period of six months); the wealthiest population quintile of about 20 percent. Poorer families are much more likely to seek help in public health centers or hospitals, while better-off families prefer to use private doctors and clinics. For many of the poor, such visits cut deeply into the family budget, since only half of them have health insurance. While the system of beneficiary identification, SISBEN (please refer to Table 4.8, note 5), is well-targeted (i.e., it reaches a much higher proportion of the population in income quintile 1 than in income quintile 5), its coverage is relatively low. Only 22 percent of the poorest population report having access to the SISBEN network, and with it, subsidized health services. Access to private and public health insurance such as ISS and EPS (Table 4.8, notes 3 and 4) is low for the poor and much higher for the better off.

The nexus between violent crime and poverty in Cali is not as strong as commonly believed, as will be discussed in Chapter 5. Table 4.8 shows that across all income quintiles, almost 20 percent of all family members are victims of violent acts, showing that all citizens are victims of Cali's violence.

Municipal programs I: expenditure distribution and access

Composition of municipal expenditures in 1997. In 1997, by far the largest share of municipal expenditures went to the treasury (*hacienda, catastro y tesoro*)—largely for servicing the municipal debt—and to building and maintaining the city's transport infrastructure. As shown in Table 4.9., these two items alone account for more than half of all municipal expenditures. Social expenditures in health, education, pacification, and social welfare have significantly lower importance in the budget.[37]

Table 4.9. Composition of municipal expenditures, 1997[38]

Secretariat/Department	Expenditure share (%)
Road infrastructure & transport	31.6
Finance *(hacienda)*	23.6
Public health	13.0
Social Housing	8.3
Education	6.6
Pacification *(convivencia)*	5.3
Social welfare	2.6
Other	9.0

Source: Departamento Administrativo de Planeación Municipal, Alcaldía de Santiago de Cali, 1999.

Geographic incidence of city expenditures. Of total expenditures (382 billion pesos or $194 million) in 1997, only 26 percent benefited specific *comunas*. The bulk of the expenditures financed either debt payments, municipal administration, or spatial planning)[39].

---

[37] In Cali, the "municipalization" of education expenditures is not complete, so part of the education expenditures are directly provided to schools by the Valle del Cauca Department. These expenditures are not included in the discussion.
[38] This table, although similar to Table 7.1, exhibits several differences: Table 4.9 refers to executed expenditures by secretariats of the municipality (each secretariat accounts for some administrative expenses), including debt service, in 1997, while Table 7.1 reports executed expenditures by program (some programs cut across different secretariats), excluding debt service, in 1998.
[39] See Chapter 7.

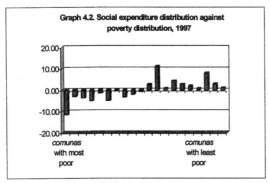 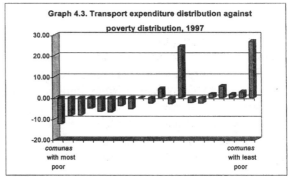

Source: EPSOC, 1999 and Municipality of Santiago de Cali, 1997.

The geographically targeted expenditures, however, went largely to the better-off *comunas*. We examine the spatial distribution of expenditures in three categories: social expenditures (social welfare, pacification, education, health, sports, public housing); transport (roads, maintenance, transit); and all other expenditures. Graphs 4.2. to 4.4. compare the distribution of these expenditures against the distribution of the poor in the city. That is, we have calculated the share of each expenditure item flowing to a specific *comuna* and subtracted from this the share of the total poor living in the same *comuna*. A positive bar will hence signal that the *comuna* has obtained a higher share of expenditures than it would have obtained had expenditures been distributed strictly according to a poverty map. In the three charts, we have ordered *comunas* from left to right by the share of the poor – the better off the *comuna*, the further it is situated to the right in the graph. As can be seen, all three charts show negative bars towards the left (the poorer *comunas*) and positive values to the right (the better-off *comunas*). Hence, the spatial distribution of expenditures in Cali was anti-poor.

Obviously, the criterion applied above to analyze the geographic distribution of city expenditures is very strict. It implies that the spatial distribution of expenditures would ideally follow the distribution of poverty in the city. For many investments, especially bulky investments, that benefit the whole city, this criterion will not be appropriate. However, the distribution viewed above has a very clear anti-poor bias in all expenditure categories, which will have to be addressed if the city is to become a more integrative and equal place for its inhabitants.

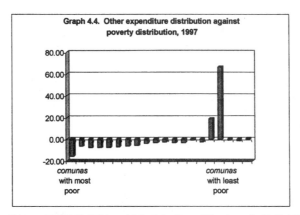

Source: EPSOC, 1999 and Municipality of Santiago de Cali, 1997.

Currently, the distribution of city expenditures perpetuates spatial inequalities. One reason for the observed anti-poor distribution of city expenditures is the formula that guides *comuna* allocations. The first component of the formula distributes 60 percent of expenditures equally across *comunas*, regardless of their population and poverty levels. The second component takes into account population, number of houses classified as strata 1 and 2, and the fiscal effort of the area. Clearly, the last two items bias against poorer *comunas* (see strata discussion in the second section).

Individual incidence of city expenditures in education and health. For a select number of programs, we can use the EPSOC survey to compute the distribution of benefits by income group. In doing so, however, we make one crucial assumption: we equate equal access with equal benefit, and hence assume that a child going to a public primary school in Aguablanca is going to obtain the same value from education as a child going to a public primary school in *comuna* 17. Obviously, this is a very stark assumption. Since it is likely that benefits per student are higher in wealthier *comunas*, our results will tend to have a pro-poor bias, which we have to take into account when interpreting them.

Most public health and education expenditures in Cali are progressive. Table 4.10 shows that the largest share of public expenditures in these sectors benefits the poorest two income quintiles the most, which is desirable. Since the poor tend to be more intensive users of public primary and secondary education, a higher portion of the benefit of these expenditures flows to them—and the same holds for public health centers and public hospitals. Although, as noted above, the coverage of SISBEN is relatively small among the poor,[40] benefits derived from being assigned to the system flow in a large proportion to the very poor—43 percent of benefits go to the poorest 20 percent of Caleños. Subsidies in higher education, however, go largely to the better-off income groups.

Table 4.10. **Incidence of education and health expenditures in Cali, September 1999** (percent)

Type of education expenditure	Income quintile					Total
	1	2	3	4	5	
Primary education	32.9	30.5	21.8	9.3	5.6	100.0
Secondary education	23.4	25.7	21.7	18.8	10.4	100.0
University education	4.2	15.3	23.2	24.5	32.8	100.0
						100.0
Public health centers	31.0	24.9	25.1	11.9	7.1	100.0
Public hospitals	27.7	22.1	24.4	12.3	13.5	100.0
SISBEN	44.2	23.7	21.2	9.3	1.6	100.0

Note: Totals may differ from 100 percent due to rounding of individual figures.
Source: Own calculations based on EPSOC, 1999.

Municipal programs II: satisfaction with basic and human service provision

Access to and distribution of city expenditures is one way to look at municipal programs—another is to see how the population rates them in terms of quality. Part of the EPSOC survey was geared to obtaining feedback from the population on satisfaction levels with municipal services in education, health, and infrastructure provision. We were particularly interested to see whether the ratings differed between the poor and non-poor populations.

As Table 4.11 shows, satisfaction levels vary significantly across services but little across income groups. Contrary to expectations, students attending school show little dissatisfaction, although a further breakdown reveals that dissatisfaction with public education (12 percent) is more than double that for private education (5 percent). Electricity, water, and garbage collection are also viewed as satisfactory by a large majority of the population, but about a quarter of the population are dissatisfied with health services . Questioned further about their negatives views of public health services, most respondents cited long waiting times as an extreme inconvenience. The cities' sewerage service gets similarly low

---

[40] Becerra, 1997, reports that focus group discussions in poor neighborhoods have also criticized SISBEN for its low coverage (p. 12).

ratings. And two-thirds of the population have especially negative views of environmental cleanliness (*saneamiento*), which has four components—regularity of: (a) fumigation campaigns (against bacteria); (b) cleaning of public parks; (c) cleaning of public canals and tubes; and (d) cleaning of neighborhoods (streets and sidewalks). Although all four aspects of such environmental cleanliness obtained negative ratings, dissatisfaction was clearly highest with the fumigation campaigns—and this was true in all parts of the city.

Table 4.11. Dissatisfaction with basic social and infrastructure services, Cali, September 1999 (percent)

Dissatisfaction with	Income quintile					Average
	1	2	3	4	5	
Education (students)	9.1	9.2	7.0	9.5	8.2	8.6
Electricity service	8.2	11.6	9.3	5.4	6.2	8.1
Water service	8.7	8.6	7.2	12.3	7.2	8.8
Garbage collection	9.2	7.7	10.1	12.8	11.2	10.2
Health (those using them) [public health services?]	24.7	16.2	17.9	16.0	17.5	18.4
Sewerage	33.8	23.1	21.9	26.0	20.1	25.0
Environmental cleanliness	61.2	66.8	60.5	64.3	60.9	62.7

Source: Own calculations based on EPSOC, 1999.

It is of particular importance for the development of anti-poverty policies to understand the priorities of the poor with respect to municipal services—and to assess whether such priorities differ from those of the non-poor population. The EPSOC survey therefore included a module to identify such priorities. The module did not purport to be comprehensive or include all possible programs; rather, it gave respondents a limited choice of basic programs. The results should be interpreted in that context. Tables 4.12 and 4.13 contain the results, again broken down by income group. Table 4.12 reports households' priorities as to the type of program (of the 13 included in the table) that should be expanded if the municipality had available resources.[41] Table 4.13 then reports on the reverse scenario, in which respondents had to choose one program for a reduction in municipal funding, if a reduction was necessary. A clear pattern as to the priorities of the population emerges from these two tables. If additional resources were available, people in all quintiles thought that education, health, employment generation, and nutrition programs should benefit first. In the reverse case (Table 4.12) these are exactly the programs that all income groups thought should be protected from cuts. Instead, the population suggests cuts in expenditures for sports arenas,[42] the police, public transport, and lighting in the face of a limited budget. All of these four areas were at the bottom of the list for expansion. These results do not mean that sports arenas, the police, and public transport are not important for the citizens of Cali, but rather that, given a constrained budget, these are not the priorities.

With few exceptions, priorities across income groups and *comunas* are very similar. Social housing and nutrition programs fare somewhat higher on the list of priorities for the poor than for better-off groups, but overall, differences in priorities are indeed small. This is a very positive result for social policymakers, as the city population is not deeply divided among different alternatives in a scarce resource environment. The results make it clear, for example, that all population groups in Cali would much prefer an expanded education and health system than an underground metro—although the latter is currently being considered by the administration.[43]

---

[41] Respondents could state only one preference.
[42] The "sports arenas" item does not capture the whole spectrum of sports activities.
[43] From a financial point of view, Solans, 1999, comes to the conclusion that the proposal for a metro is not viable.

Proposed strategic priorities for social development in Cali

Cali is not short of strategic plans to reformulate its social policy. Quite the contrary. First, since the beginning of the 1990s, the city has been a *Municipalidad Saludable por la Paz*, adopting the World Health Organization's concept of a healthy city. This is a development plan that goes significantly beyond health sector planning and interventions, to strive for inclusive city development. Two recent municipal decrees have reaffirmed Cali's aspirations to be a healthy city, but integrative planning along these lines has not taken place to date.

Table 4.12. Priorities for expansion of municipal programs, Cali, September 1999 (percent)

Program	Income quintile					Average
	1	2	3	4	5	
Education	31.3	30.9	29.2	32.3	34.8	31.7
Health	19.5	19.9	30.2	23.6	23.9	23.4
Employment & income prog.	18.9	22.2	18.6	18.7	19.8	19.7
Nutrition programs	8.8	4.4	5.6	5.6	1.2	5.1
Social housing	10.4	11.7	8.4	5.9	5.5	4.8
Police	3.2	3.1	2.6	7.2	7.8	4.5
Water	2.1	1.5	1.5	1.9	1.4	1.7
Electric lighting	1.8	1.8	0.8	0.5	1.5	1.3
Communal households (ICBF)	1.0	2.6	0.6	1.5	0.6	1.2
Public transport and roads	1.4	0.8	0.9	0.8	2.0	1.2
Sports arenas	1.3	0.9	1.3	0.7	0.8	1.0
Sewerage	0.5	0.1	0.2	0.8	0.2	0.4
Garbage collection	0.0	0.0	0.1	0.7	0.5	0.3

Source: EPSOC, 1999.

Table 4.13. Priorities for cut-back of municipal programs, Cali, September 1999 (percent)

Program	Income quintile					Average
	1	2	3	4	5	
Health	0.9	0.2	0.4	0.8	0.5	0.6
Education	0.8	0.6	1.1	0.7	1.1	0.9
Water	2.8	1.0	1.1	0.7	0.3	1.2
Nutrition programs	0.9	2.5	0.8	3.3	1.8	1.9
Empl. & income programs	0.5	1.7	3.1	1.9	3.6	2.2
Garbage collection	0.7	3.8	1.9	1.6	2.6	2.2
Sewerage	1.1	3.3	4.2	3.2	2.1	2.8
Social housing	3.3	1.1	5.5	5.5	3.7	3.8
Communal households (ICBF)	10.5	9.5	5.5	5.9	8.0	7.9
Electric lighting	7.6	14.1	12.5	10.1	9.4	10.8
Public transport	18.2	12.9	12.9	17.5	18.0	15.9
Police	18.6	17.2	16.6	15.1	12.9	16.1
Sports arenas	33.9	32.1	34.3	33.6	35.9	33.9

Source: EPSOC, 1999.

Second, the city's development plan (1998 to 2001) sees social investment and citizens' participation as priorities in the local management agenda, with the explicit objective of achieving equal access to opportunities—which are seen as integral to the development of the population. Departmental plans for the period 1998 to 2001 set out specific targets and goals for education, health, housing, and social welfare. Most of these targets are designed to close existing deficits in the supply of key services, and to promote the active participation of the population in program design.

Third, in parallel with publishing the city development plan at the beginning of its term in office, the current municipal administration established the *Comisión Asesora de Politica Social y Gestión Comunitaria* as a consultative body on social policy formulation; this multi-partite body included members of the non-governmental sector, private sector, municipal government, and academia. The administration also charged the *Foro Nacional por Colombia* with coordinating formulation of a permanent social policy for Cali. The *Comisión's* report, *En Busca de la Equidad,* offers a detailed vision of a new social policy in Cali—one that is integrative, selective, strategic, and equality enhancing. *En Busca de la Equidad* recommended an anti-poverty strategy built around (a) economic reactivation; (b) improvement of city livelihood; (c) formation of human capital; and (d) formation of social capital. The document outlines dozens of sensible, specific, and innovative program and reform proposals including all areas of city policymaking and all the main actors—the private sector, the voluntary sector, academia, the municipality and citizens' groups. In view of that report, this chapter does not offer specific program suggestions. However, we believe that initiatives should address the following five general points:

- *The currently scarce and limited resources of the city need to be distributed differently, both by function and by geographic area.* The geographic distribution of expenditures is regressive for social, transport, and other expenditures. Further, the share of social expenditures in the city budget is very low. Given the low level of social spending per capita, near-term investment plans have to be re-evaluated. For example, the EPSOC survey clearly established that demand for a new metro system in Cali, in the context of a limited budget, is very low. Although a large part of the metro would be financed by the national government and by a gas surcharge that could not be spent on any other activity, the municipal government would have to raise new debt on international markets, and which would create a long-term burden on the muncipal budget.

- *The many diverse municipal programs in the social area need to be reduced in number, more centrally administered, and brought under a limited number of strategic objectives.* The collaborative process of formulating the City Development Strategy showed how many actors in social policymaking there are on the municipal side—and that a common strategy and leadership is needed to bring the diverse programs and initiatives together. PROCALI might play an important role helping the municipality in this endeavor, as it has experience in coordinating the actions and programs of numerous NGOs. It is also important to underscore that selected priorities should be given not only a prominent place in the municipal development plan, but, more importantly, a corresponding share of the municipal budget.

- *A nutritional safety net might need to be developed.* The EPSOC survey revealed a high degree of food insecurity for the poorest of the poor. At the same time, access to nutrition programs in the city is very low. Nutritional programs would, therefore, need to be high on the agenda for the municipality—and many successful examples of nutrition programs exist in Latin America. Such programs often achieve a wide reach and the desired nutritional impact if they are developed in collaboration with other programs—e.g., mother-child health care—on which they can successfully piggyback to identify beneficiaries. Non-governmental organizations are often key implementing units. Before plans are drawn up, however, an in-depth qualitative investigation would be needed to pinpoint the extent and location of food insecurity.

- *The city would benefit significantly from a permanent poverty monitoring unit.* A poverty monitoring unit, perhaps sponsored by private business and with representatives of civil society, could be used to trace food security and poverty levels in the city, and also to evaluate the impact of the city's social policies and evaluate satisfaction with municipal programs. An instrument such as the EPSOC survey could be useful for this purpose. Further, the monitoring unit could record, systemize, and publish institutional maps of the city (as are currently being developed by the Universidad del Valle), which would allow for more precise knowledge about existing programs, including their objectives, target groups, implementing agencies, and expenditures.

- *To jumpstart new initiatives, especially in employment creation and education, formal private-public partnerships may need to be created.* Given the social crisis in Cali and the municipality's budgetary situation, a short-term possibility for raising funds would be for the private and public sectors to enter into formal partnerships. Ideas for such partnerships are not lacking; *Foro por Colombia* and several other agencies have proposed, among other things, an investment fund for small-business creation, internships and training programs for the unemployed, an education fund to increase secondary enrollment in the poorest areas of the city, and a private-public partnership to reorient curriculum in secondary schools. What keeps many of these initiatives from taking shape is lack of trust among the various actors— and this is the hardest obstacle to overcome in Cali.

Chapter Annex A. Regression Result for the Wage Regression (OLS)

OLS Regression

Dependent variable: log of hourly wages
R-squared (adjusted) : .44
f-test: 42.6

Variable	Parameter	t-statistic	description
INTERCEP	6.147383	43.062	intercept
E4SEXO	0.172500	4.992	sex (dummy variable)
E5EDAD	0.053177	8.155	age of individual
AGESQ	-0.000597	-7.443	age squared of individual
INPRIM	-0.370479	-6.102	incomplete primary education, dummy variable
PRIM	-0.094675	-1.778	complete primary education (dummy variable)
INSEC	-0.158696	-2.321	incomplete secondary education (dummy)
SECOND	0.092982	2.002	complete secondary education (dummy)
SUPERIOR	0.463618	8.151	superior education (dummy)
MARRIED	0.136645	3.769	married (dummy variable)
G4_0	0.336143	3.877	occupational category 0 (dummy)
G4_1	0.303989	3.798	occupational category 1 (dummy)
G4_2	0.674272	4.540	occupational category 2 (dummy)
G4_3	0.085143	1.284	occupational category 3 (dummy)
G4_4	-0.009804	-0.183	occupational category 4 (dummy)
G4_5	-0.018645	-0.326	occupational category 5 (dummy)
G4_6	0.139789	0.981	occupational category 6 (dummy)
G4_7	0.003868	0.051	occupational category 7 (dummy)
G4_8	-0.055254	-0.846	occupational category 8 (dummy)
G6ROL_2	0.288328	3.735	employer (dummy)
G6ROL_3	-0.104105	-1.361	public sector employee (dummy)
G6ROL_4	-0.101520	-2.146	private sector employee (dummy)
G6ROL_5	0.068433	0.830	contract worker (dummy)
G6ROL_6	-0.325246	-4.350	domestic employee (dummy)
G6ROL_7	-0.749738	-1.175	apprentice
ESCRI_1	0.097233	2.063	contract of employment exists (dummy)
EXPMONTH	0.000609	3.210	job experience (in month)
SIN	0.181218	1.961	individual with union affiliation
PEN	0.153251	3.692	individual paying social security (pension) insurance at work
REGNOOR	-0.296155	6.190	area dummy variable (North)
REGSUOR	-0.337167	6.937	area dummy variable (South-east)
REGLAD	-0.298064	5.280	area dummy variable (ladera)
REGMIT1	-0.243581	5.193	area dummy variable (north-east and central)

Source: EPSOC, 1999.

5. URBAN VIOLENCE

Introduction

Violence is a serious and widespread problem in Colombia, and especially in Cali. It is a complex and multifaceted phenomenon that affects the daily lives of all Colombians. Given its intricate nature, effective urban violence prevention programs must involve multiple levels and sectors simultaneously, making such programs difficult to coordinate and sustain. Despite the disheartening situation, there is hope for the country and the city. Since the early 1990s, some urban violence reduction policies have been successful in Colombia's largest cities, including Cali, and further positive results could be obtained with appropriate strategic programs and interventions.

This chapter is organized in six sections. The following section describes the national context and historical background to the current violence crisis. The third section is a diagnosis of the specific violence manifestations in Cali, based on local crime statistics. The fourth section explores the relationship between violence, particularly homicide rates, and key social, institutional, political, and legal factors in Cali. The fifth section presents an overview of current violence reduction programs in Cali, and the final section lays out the proposed strategic priorities, as they were discussed with local counterparts in September 1999, during development of the second stage of the CDS.

The national context

Since the early 1980s, Colombia's domestic security situation has grown increasingly worse due to rising levels of organized crime by drug cartels, guerrilla groups, and paramilitary organizations—a problem made worse by absence of strong law and order instruments and high levels of impunity.

This is not Colombia's first violence crisis. Colombia suffered a major domestic breakdown during the 1940s and 1950s, in the wake of a civil war between the two main political parties, Liberals and Conservatives. During this period, known as *La Violencia*, an estimated 200,000 people were killed. *La Violencia* came to an end when the parties agreed on administrative and political power sharing. This period, which lasted 16 years (1958-1974), was known as the National Front.

During the early years of the National Front, political violence decreased significantly, and it was generally agreed that there was no need for major institutional change. However, during the 1960s and 1970s, new forms of violence appeared, particularly small Communist guerrilla group operations in rural areas, and common delinquency in urban areas. This violence was met with poor institutional performance, including high levels of impunity and extremely inefficient and often corrupt law enforcement agencies. The probability that any one delinquent act would lead to sentencing was around 10 percent in the 1960s, and this decreased to 5 percent during the 1970s.[44]

By the end of the 1970s, a new phenomenon, illegal drug trafficking, added another dimension to Colombian violence. Paramilitary groups boomed, originally as a protection service for drug lords. A variety of peace initiatives with both guerrilla groups and drug cartels were undertaken during the 1980s and led to mixed results; but until the early 1990s, no serious programs were undertaken to strengthen law enforcement or define strategic approaches for violence and crime reduction. By the time such programs were undertaken, violence had already reached unprecedented levels, characterized by car bombs and other acts of terror in cities, and by kidnappings and massacres in rural areas. The direct and indirect costs of the crisis, including high levels of political and administrative corruption, rapidly became apparent.

---

[44] Rubio, 1996.

International pressure, mainly from the United States, forced the Colombian police and intelligence agencies to develop more efficient counter-narcotic strategies to fight organized crime. This led to the relatively successful break- up of various important drug cartels, particularly in Medellin (1991-93) and in Cali (1994-95). At the same time, the new Constitution (1991) strengthened political, financial, and administrative decentralization, and increased local administrative responsibility for crime and violence issues (though the police department continues to be a national agency). These two factors led to a change in the way violence was perceived and handled. Since the early 1990s, Cali, Medellin, and Bogotá have introduced similar strategies of urban violence reduction, based on a combination of law enforcement, risk factor prevention, and public information campaigns. It is generally believed that these campaigns have contributed to the significant decrease of urban violence in these cities over the past few years.[45] While the national homicide rate decreased some 5 percent between 1994-1998, Bogotá, Medellin, and Cali experienced reductions of 27 percent, 35 percent, and 27 percent, respectively. In general terms, the initiatives were related to a wider strategy of "bringing in the state," in recognition of the fact that Colombia needs not only a political settlement of the internal conflict, but also efficient law enforcement institutions.

Crime and violence in Cali

Available statistical information on violence, particularly homicides, is extremely good in Cali in comparison with other cities in Latin America. These statistics show that, since 1990, common crime has consistently been lower in Cali than in Colombia, whereas homicide rates have been higher, particularly in recent years. This phenomenon is the result of the dynamics of violence in the city and presents very distinctive characteristics in terms of spatial occurrence, and demographic and socioeconomic profiles of the victims.

Table 5.1. presents the <u>reported crime rates</u> for the period 1990-1998 for Colombia and the metropolitan area of Cali. The city has maintained lower rates than the country as a whole. For the Cali metropolitan area, crime rates have been slowly but steadily decreasing since 1994, a result that has not been adequately explained. In terms of homicides, the situation is quite different: Cali has had higher rates than the country as a whole since 1992. Between 1990 and 1994, the homicide rate increased in Cali and then decreased in the period 1994-1997. In 1998 there is a slight increase, and preliminary data indicate that during 1999 the rate increased to 95 per 100,000. Given that homicides have been, and continue to be, the worst violence problem in Cali, this chapter will focus mainly on that issue.

Peaks in the homicide rate in metropolitan Cali (please refer to Graph 5.1) can be explained by the dynamic aspects of violence in the city. During the early 1980s, when the homicide rate oscillated around 40 per 100,000, violence was characterized by urban guerrilla penetration in certain parts of the city, particularly the Siloe neighborhood (in *Comuna* 20), and confrontations between guerrilla and public force inside the city. Later in the decade, urban guerrilla networks disintegrated but other factors appeared: (a) local penetration of crime related to drug trafficking; (b) relatively high levels of more disorganized forms of violence; and (c) social cleansing campaigns by public forces and private actors.[46] The homicide during this period averaged 60 per 100,000. Then came the extremely violent first part of the 1990s, marked by mafia-style executions, which should be understood in the context of a public crackdown on the local drug cartel, the turf wars among criminal organizations, and the strengthening of other guerrilla groups in and outside the city.

---

[45] Castro, 1997.
[46] Camacho, 1992.

Table 5.1. Reported crime and homicide rates in Colombia and the Cali metropolitan area, 1990-1998

Year	Crime rates		Homicide rates	
	Colombia	Cali, metro area	Colombia	Cali, metro area
1990	574	485	75	60
1991	605	486	86	77
1992	600	499	84	85
1993	501	463	83	90
1994	556	492	78	119
1995	575	489	72	107
1996	587	448	76	101
1997	577	399	68	81
1998	N/A	361	57	83
1999	500	400	59	95

Notes: Metropolitan area includes Cali, Yumbo, Palmira, La Candelaria, and Florida.
Crime rate = number of crimes/100,000 inhabitants.
Homicide rate = number of homicides/100,000 inhabitants.
Sources: Urrea and Ortiz, 1999; Franco, 1999; *Policia Nacional,* 2000.

The homicide rate decreased during the 1995-97 period, due to a combination of factors: (a) partial neutralization of major local drug trafficking organizations; (b) improved local policies of violence reduction[47]; and (c) improved performance of national law enforcement agencies, particularly the police and the justice administration. The homicide rate then increased again in 1999, due to: (a) confrontations between guerrillas and paramilitary organizations, have strengthened their presence both in the Valle del Cauca Department and inside certain parts of Cali; (b) confrontations between guerilla and government forces; (c) a rise in common crime, partly as a result of the socioeconomic crisis; and (d) diminished effectiveness of local policies of violence reduction, as a consequence of the local political and financial crisis.

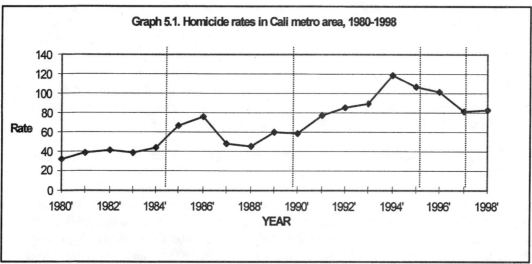

Source: Urrea and Ortiz, 1999.

---

[47] For an evaluation of national strategies of violence reduction, see Deas, 1999.

Homicides do not affect all areas of Cali in the same way; homicide mapping shows clear geographic patterns. Although the ranking of individual *comunas* varied between 1997 and 1999, the following areas are generally identified as the most violent parts of the city (please refer to Map 5.1):

- The City Center—*comuna* 9 and particularly *comuna* 3. The latter includes *la Olla,* a section with a high concentration of homelessness, prostitution, drug dealing, petty delinquency, and an ongoing turf war between two well-known and extremely violent gangs.

- The Northwest, known as *Ladera*—*comuna* 1[48] and particularly *comuna* 20. The latter includes *barrio* Siloe, a neighborhood where violence is associated with a large number of gangs that may operate under influence of guerrillas (*milicias*) or organized crime (*sicariato*), or in a more autonomous way (*pandillas*).

- The Southeast—especially *comunas* 11, 12, 14, and 7. These correspond in part to the city district of *Aguablanca.* Homicides are attributed to interpersonal violence, to gangs and their territorial conflicts, and to penetration by armed groups, although *milicias* seem to be less powerful than in Ladera.

- Rural area.. High homicide rates in the extended but sparsely populated (30,000 population) rural part of Cali seems to be related to the presence of guerrilla groups and organized crime.

Map 5.1. Homicide rates per *comuna* in Cali, 1997

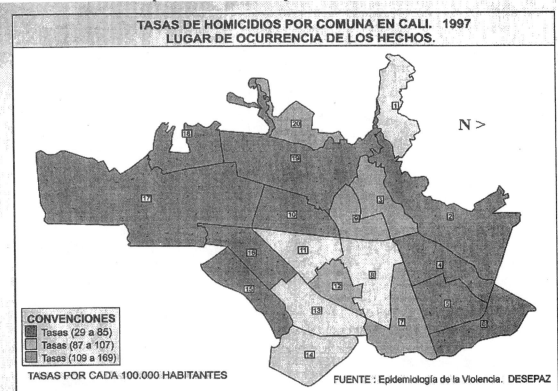

Source: DESEPAZ, 1998.

---

[48] *Comuna* 1 had a high rate of 93 per 100,000 in 1997, but the rate decreased to 65 per 100,000 in 1998 and to 67 per 100,000 in 1999. Therefore, according to the most recent data, it is not one of the most violent *comunas*. All other *comunas* ranked as very violent in 1997 remained in that category in 1998 and 1999.

Cali has more than 300 neighborhoods and some 15 per *comuna*. Inside each *comuna*, significant differences in homicide rates and types of violence exist per neighborhood, and even per street or block. Thirteen of the more than 300 neighborhoods accounted for over 20 percent of total homicides in 1996 (please refer to Table 5.2.). Crime data show that particular neighborhoods can undergo significant increases or decreases in homicide rates from one year to the next, and even over shorter periods. Such fluctuations mainly reflect changes in the operational logic of organized crime or the operational behavior of law enforcement agencies.

The residence of homicide victims is also geographically concentrated, although less so than the related crime scenes.[49] Since impunity levels are extremely high for homicides, no significant data are available on the residence of victimizers. It is generally believed that they live in the neighborhoods with the highest incidence of homicides.

Table 5.2. Contribution to Cali's homicides by most violent neighborhoods, 1996

Neighborhood	*Comuna*	Contribution to Cali's homicides (percent)
Siloe	3	3.4
Sucre	4	2.4
Manuela Beltran	11	1.8
Marroquin I	12	1.8
Alfonso Bonilla	14	1.5
El Calvario	3	1.4
Santa Elena	10	1.3
Obrero	9	1.3
Mojica I	15	1.2
El Rodeo	12	1.2
El Retiro	15	1.1
Antonio Nariño	16	1.1
Terron Colorado	1	1.1
TOTAL		20.6

Source: Guzman and Dominguez, 1996.

In terms of <u>demographic characteristics</u>, homicide victims tend to be mostly young men. In 42.6 percent of all homicide cases, the victims range in age from 15 to 25. Even more overwhelming is the fact that 93.2 percent of all homicide victims are males. One manifestation of violence in the young population is the proliferation of gangs. More than 100 street gangs, made up mainly of adolescents and young adults, are estimated to be active in different neighborhoods of Cali. Some gangs are extremely violent, particularly those that have proliferated in the shadow of drug trafficking and other organized criminal activities. Youngsters belonging to these gangs tend to trivialize violence, a phenomenon expressed in a short-term vision of life and a lack of interest in education, work, and health issues such as HIV/AIDS. There is no systematic analysis available on gangs in Cali in terms of age structure, amount of members, degree of implication in crime, and drug consumption, which leads to dangerous stereotyping.

Street kids (*gamines*) constitute another group severely affected by homicides. Cali has an estimated 800 to 1,200 street kids that live permanently in the streets. An additional 10,000 children are estimated to

---

[49] In 25 percent of the cases, no residence was established for the homicide victim.

"walk the streets." Children of both groups can be involved in drug dealing, prostitution, petty delinquency, and robbery. They can also be homicide victims and the object of social cleansing operations.

In terms of socioeconomic characteristics, there is little quantitative data about the homicide victims, other than the following: (a) victims tend to be workers in construction, commerce, and informal sectors; (b) in 36 percent of cases, the victim had a high level of alcohol in the blood;[50] and (c) 20 percent of the victims had been killed after being robbed, while 14 percent had engaged in fights that ended fatally.[51,52] In addition to these characteristics, place of homicide occurrence, although not a perfect indicator, helps in understanding who the victims are. Graph 5.2 ranks *comunas* by homicide rate and by poverty rate. Poverty rates come from the EPSOC household survey and are explained in more detail in Chapter 4. The table shows that, although there is some coincidence between the two rankings, it is far from a perfect correlation. *Comunas* 3, 9, 12, and 11, all quite violent, are not in the group of poorest *comunas*. Conversely, *comunas* 16, 1, 18, and 6, which are poor, do not present high levels of homicides. Two observations can be made. First, socioeconomic heterogeneity within *comunas* is large. There are pockets of extreme poverty in specific *comunas* that overall do not appear to have a high level of poverty. Crime and violence also vary strongly among neighborhoods within the same *comuna*. Second, violence in Cali is a multi-dimensional problem, with different causes and dynamics in different parts of the city. The intensity of the violence in *comuna* 20, for example, is related to issues of organized crime and guerrilla penetration in a peripheral part of the city, rather than to issues of poverty. *Comunas* 3 and 9 are

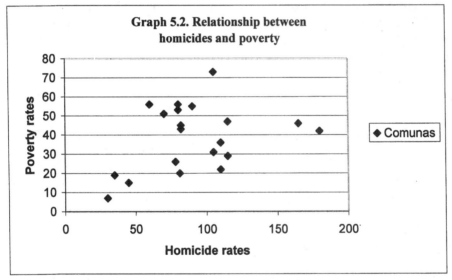

Source: Hentschel, Mehra, and Seshagiri, 1999 and Desepaz 1999.

in central parts of the city, with commercial areas but also with high levels of prostitution, street children, drug dealing and drug consumption, illegal activities that generate crime and violence. These areas have been at the center of violent competition between gangs. In addition, as shown in Table 4.8, in all income quintiles the percentage of households having a family member who has been a victim of violence is a more or less constant 20 percent. Most likely the type of violence is different in each income quintile, but this was not captured in the EPSOC survey. The conclusion is that the nexus between poverty and violence is still unclear and that further research is needed.

---

[50] Urrea and Ortiz, 1999.
[51] For 59 percent of the cases, the causes of the homicide are unspecified.
[52] Municipal Program for Development, Security and Peace (DESEPAZ), 1999.

Relationship between violence and social, institutional, political, and legal aspects

Violence in Cali is associated with a complex mix of interpersonal, social, economic, and structural factors. It is carried out by a variety of armed protagonists who range from the rather disorganized to the extremely well organized, and who operate for motives that range from prosaic to political and ideological. Violence in Cali has undergone significant change over the last 15 years, and has become, to a certain extent, the context in which the society has evolved. Although violence in Cali has significantly decreased between 1994-1997, this does not seem to have had an impact on perceptions of insecurity. A 1998 survey showed that 61 percent of Cali residents were very afraid of becoming a victim of violence, 22 percent somewhat afraid, and only 14 percent not afraid at all.[53] In this section the following aspects, and their relationship with violence in the city, are reviewed: social, institutional, and political crises, and weak law enforcement. It is difficult to separate these aspects as causes or consequences, since violence fosters the destruction of social, political, human, economic and physical resources, which in turn fosters more violence.

Over the last decades, in the context of rapid and disorganized urban growth, Cali has experienced a process of <u>individualization</u> and secularization, as is typical of a modernizing society. Individualization—along with the city's social, economic, and leadership crises—has weakened traditional forms of social control at the local level (please refer to Box 5.1). It has also increased mistrust in elected officials, public agencies, and the police, and among the population. This mistrust translates into apathy, cynicism, and a lack of participation. Although Caleños like to remember their city as Colombia's capital of civility, people now say that life in the city is characterized by "weakening of citizen participation," "deterioration of social norms and networks of cooperation," "lack of social trust among citizens as well as between citizens and the city administration," "lack of tolerance," and "increasing individualism"—all of which reflect a crisis of social capital.

Box 5.1. Deterioration of networks of cooperation

"Last night they robbed me. They stole my shoes. There were a lot of people who saw it, but nobody got involved. Nobody wanted to show that they were interested in my problem. So…I got robbed in front of everybody. It is always like that. When somebody else gets robbed and you happen to see it, you do not get involved either. Since the other is not interested in what happens to you, you are not interested in what happens to the other. Since nobody is interested in anybody here, nobody is interested in what happens to anyone else."

Source: Youngster, City District of Aguablanca; in Vanegas Muñoz, 1998, p. 169, translated by Gerard Martin.

The process of modernization has not been accompanied by the creation of dynamic and inclusive public institutions or a common social vison. Instead, the construction of local society came to be dominated by particular short-term interests. To make things worse, the effects of organized crime on Cali's social texture have been pervasive, creating <u>high levels of corruption and an unprecedented breakdown of trust among the local political elite</u>, resulting in a crisis of the institutional apparatus. Therefore, in contradiction to the view that violence is simply a result of a deteriorating social fabric, citizens point to bad governance as the main problem. They complain that local and national government has not been strongly committed to violence reduction, while public support has been rather generous, particularly through peace marches and violence reduction campaigns.

---

[53] Atehortúa et al, 1998; Moser and McIlwaine, 1999.

Political crises have also induced violence in Cali. There are several dimensions to this, particularly political intolerance, the presence of guerrilla and paramilitary groups, and the migration of internal refugees toward Cali. Political intolerance, including physical elimination of political opponents (or their supporters), co-option by clientelist networks, and forced obedience to political chiefs have a longstanding tradition in Colombia, and Valle del Cauca is no exception.[54] While these practices were brought largely (but not completely) to an end during the National Front regime, political parties and their leaders still need to formally distance themselves from these practices. Although the new Constitution (1991) and subsequent electoral reforms have improved conditions for participation, political intolerance and coercion continues to deprive many Colombians and Caleños in particular, of their full rights as citizens.

Guerrilla groups operate in the urban periphery of Cali, seriously disrupting social and economic activities through kidnappings, truck stealing, imposition of war taxes, and selective killings. Various police posts in peripheral parts of Cali have been the object, over the last years, of guerrilla assaults. Part of the drug trade in rural areas of Valle del Cauca has supposedly been taken over by the guerrillas. The roads from Cali to Buenaventura and the Pan-American highway to Ecuador are highly insecure, as a consequence of guerrilla operations. There is evidence of recent strengthening of paramilitary groups in the region and of their intention to "re-conquer" the roads to Buenaventura and Ecuador. Such operations would imply high levels of violence in Cali's rural hinterland, with potential repercussions inside Cali. Urban guerrillas, called *milicias,* operate in certain parts of the city, particularly in the Ladera section, where they have claim to have introduced "popular justice" and safety in neighborhoods. In fact, neighborhoods where *milicias* are present are consitently among the most violent in the city. Areas of the city plagued by *milicias,* gangs, and paramilitary forces are fragmented into micro-territories.[55]

An estimated 75,000 internal refugees, fleeing zones with even higher levels of violent conflict, have settled in Cali since 1985, including a record 13,325 new refugees in 1998.[56] More than two thirds of these intend to stay in Cali, though the city has no special centers to lodge them. Some neighborhoods harbor numerically significant groups of refugees, a situation that could generate social tensions and further violence.

Box 5.2. Mistrust between the police and the community

"*Vea*, in these neighborhoods reigns the law of silence. The majority of people don't say anything because they're afraid they will get killed. I have been working in this sector for some months now but I don't know who the killers are. Well, I have heard that X is the killer in the neighborhoods El Retiro and El Vergel. People know the killers, and they know that there are confrontations between groups and youth gangs. But although the people know them, they won't tell me who they are. People are being killed but nobody says anything. As if they prefer to resolve problems here by killing each other. There are many people here who are interested in resolving this by violence. Nobody here is interested in justice. As if this will only stop when everybody who has got to be killed will have been killed."

Source: Police agent, City District of Aguablanca; Vanegas Muñoz, 1998, p. 112), translated by Gerard Martin.

Weak law enforcement, impunity, and human rights violations are central to the spread of violence. The police continue to operate under extremely difficult circumstances: (a) human and infrastructure resources have improved but are still insufficient; (b) a high rate of rotation among top officers leads to short institutional memory; (c) various police stations inside Cali's urban perimeter have been assaulted by armed groups over the last years; and (d) infiltration of the police force by organized crime is a permanent

---

[54] Atehortúa, 1995.
[55] Vanegas Muñoz, 1998.
[56] According to UNICEF

risk. In addition, (v) while their image has improved, the police are still perceived as corrupt, inefficient, arbitrary, and unwilling or unable to enforce the law (please refer to Box 5.2). These conditions are fertile ground for crime, private justice, retaliation, and vengeance as well as practices of social cleansing.

Graph 5.3. Perception of police performance, 1999

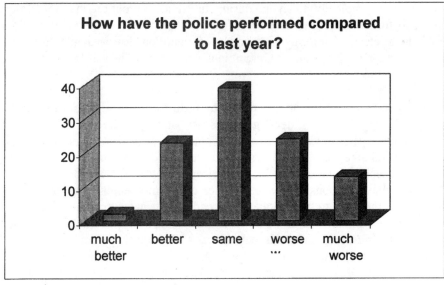

Source: EPSOC, 1999.

A major problem in Cali, as in the rest of Colombia, is the juvenile justice system. Juvenile offenders cannot, under any circumstance, be sent to an adult prison, but youth prisons as such do not exist. Consequences of this situation are: (a) tension and lack of trust between the justice administration and the police, since sentenced juvenile criminals are often quickly back in their neighborhoods; (b) lack of public trust in the justice administration; (c) tacit popular support for popular justice and social cleansing (please refer to paragraph 5.27). While the municipal administration and others have recognized this situation, no structural solutions have been proposed, and will depend mainly on new national legislation. Cali has only one juvenile rehabilitation center, Valle de Lilli. The presence of many extremely violent youth in the center has created a highly dangerous situation. Armed groups have entered Valle de Lili to liberate their comrades or to execute their enemies. Another problem is that Valle de Lili has only 400 places to serve not only Cali (with its 1,800 homicides a year, some committed by juveniles), but the whole of Valle de Cauca, as well as surrounding departments.

In addition to the crisis in the juvenile justice system, Valle del Cauca and Cali experience, like the rest of Colombia, are experiencing a critical penitentiary situation, not only in terms of overcrowding (in part as a result of increased prosecution) and corruption (e.g., in terms of parole and other benefits), but also in deficient health and security conditions and rehabilitation services. There is also a lack of temporary retention centers, so those arrested are held in an open-air facility on the street in front of the Attorney General's office, in the center of the city.

Practices of social cleansing take place in Cali. No precise data are available, since corpses of victims are seldom reclaimed, and no judiciary procedures initiated. Three types of victims have been identified in order of importance: (a) (supposed) delinquents, criminals, and hit boys (*sicarios*); (b) drug addicts and street corner drug dealers; (c) prostitutes, street kids, the homeless, and other marginal people. Social cleansing has been attributed to a variety of actors, particularly urban guerrillas, death squads, public forces, and street gangs, who often legitimize their operations in terms of popular justice: protection of citizens by elimination of "non-citizens." Social cleansing may receive passive popular support in crime-ridden neighborhoods, given the high levels of impunity and the readiness and availability of armed actors to do the job.

Current programs for violence reduction in Cali

There are a great number and variety of programs to reduce violence in Cali. This section describes the most important in some detail. The first to attract national interest was the Municipal Program for Development, Security and Peace (DESEPAZ), introduced under the 1991-1994 administration. The program, modeled on a public health approach, focused on preventive measures, particularly control of risk factors, public information campaigns about risk factors, and promotion of civic values and community initiatives to reduce violence. Subsequent administrations have maintained the program's preventive approach, while giving more attention to law enforcement issues. In addition, DESEPAZ now operates as an advisory unit for the mayor on issues of security and peaceful coexistence.

The National Department of Planning (DNP) has developed, in coordination with other relevant agencies, and with assistance from the Inter-American Development Bank (IDB), a National Program for Peaceful Coexistence and Citizen Security (1998-2002). The program aims to improve crime data, strengthen law enforcement agencies, and promote public information on violence and crime. It has three local components (each of about $10 million), for Bogotá, Cali, and Medellin. The program is currently coordinated on the national level by the Presidential Council for Security and Coexistence, and on the local level by municipal administrations. Cali uses the IDB program as an integral part of its recently formulated Municipal Strategy for Security and Peace (MSSP), which covers the period 1999-2002. The IDB/MSSP program foresees 20 different interventions, which are or will be implemented by the Secretary of the Interior (9 projects), the Secretary of Social Development (2), the Secretary of Education (1), the Secretary of Transportation (1), DESEPAZ (4), the Metropolitan Police Department (3), and others. Serious obstacles have arisen during the first year of implementation of the IDB/MSSP program, related to a combination of factors: (a) problems with disbursement procedures; (b) the financial crisis of the municipality; (c) fragmentation of the program in a range of smaller projects that lack strong coordination; and (d) changes in strategic priorities of urban violence reduction programs following the installation of new national and local governments.

Another program, run by Cali's Chamber of Commerce, is the Commission for Citizen Coexistence, which works with private and public money. In operation since 1996, the Commission has been a crucial actor in reducing violence and promoting coexistence in Cali. It continuously lobbies on issues such as improved crime data analysis, strengthening of inter-institutional coordination, creation of public-private partnerships, violence prevention, stricter enforcement of traffic norms, rehabilitiation of public space, technical modernization of the police and the regional Office of the Attorney General, and the creation of a special unit to fight organized crime. The Commission has undertaken important studies on these issues and has financially supported a variety of initiatives. It also publishes a bi-yearly journal, *El Ojo Avizor: El Pulso de la Convivencia*, which contains basic information on violence and crime trends in Cali and Valle del Cauca. The Commission is now in the process of creating a small think tank (CORPOPAZ) and an electronic information network (OSIJ) on related issues.

The Archdioceses of Cali has been extremely supportive, in a variety of ways, of public and private initiatives to diminish violence and strengthen peaceful coexistence and tolerance. It has undertaken pioneering work on the issue of internal refugees in Cali, and on human right violations. Priests and missionaries are among the very few who operate small grassroots projects and initiatives for violence reduction in the most conflict-torn neighborhoods. Less is known on similar work by the many but less organized Protestant churches in Cali.A variety of NGOs, including nationally known organizations such as the Carvajal Foundation, actively work to prevent violence and promote peaceful coexistence. There is, however, no systematic overview available of NGOs working on issues of violence reduction or of lessons learned from their efforts. University-based research, particularly at CIDSE, CISALVA, and the regional chapter of *Foro por Colombia,* have significantly contributed to a better understanding of violence in Cali.

Table 5.3. Summary of existing programs for violence reduction in Cali

	INSTRUMENT	FUNCTION	STATUS	OBSTACLES	CURRENT POLICY INITIATIVES
Strengthening inter-institutional coordination	Unidad Coordinadora Municipal (UCM)	Coordination of components of IDB/MSSP	Nonexistent; function now fulfilled by DESEPAZ		Creation foreseen by IDB/MSSP
	DESEPAZ	Advice, design, impl and exec of relevant projects	Operative since 1992	High rotation of directors	
	Consejo Tecnico de Seguridad (CTS)	Weekly security meeting at Interior Secret	Operative		
	Concejo Mpal de Seguridad (CMS)	Mayor calls, on ad-hoc basis, for security meeting	Operative	Irregular meetings	
	Comite de Seguridad y Conv (CSC)	More permanent and participatory than CMS	Nonexistent		Creation foreseen by IDB/MSSP
	Veeduria Ciudadana	Civic accountab of Mpal Strat for Security & Peace	Nonexistent	Function in part foreseen by CSC	Promotion foreseen by IDB/MSSP
Improving crime data collection and diagnosis	Centro Op. de Seguimiento al Delito (COSED) at Office of Mpal Secr. of Interior.	Gather crime data, with input from several centers	In process of creation	Lack of (a) resources of contributing instit, (b) system data diagnosis, (c) analytical pieces	COSED is a component of IDB/MSSP, but no funding for contributing agencies
Strengthening preventive measures and controling risk factors	Plan de desarme	Prohibition to wear legal arms on particular days and areas	Intermittently applied since 1992	Non-systematic application during current administration	Will be strengthened under new natl guidelines
	Ley Zanahoria	A closing time for nightclubs at 1 AM.	Operative	(a) Opposed by night club owners, (b) modified by Mayor, leading to discredit	Will be maintained
	Curfew for juveniles	Curfews for juveniles on particular days and city sectors	Operative	Ad-hoc application	Will be maintained
	Measures to diminish violent death in traffic	Ex: mandatory wearing of moto helmets	Impl by Secr of Traffic & Police Dept	Many car drivers without legal driver's licence	Harmonic Transit Progra,
	Public information campaigns on risk factors	Public info. campaigns to promote respect, tolerance	The Mpal Adm. & Chamber of Commerce, archdioceses, NGOs	Initiatives are now ad hoc.	IDB/MSSP component
	School programs for education of tolerance and other	Teach children and youngsters on tolerance and other	Some ad hoc programs and initiatives at certain		As IDB/MSSP components, schools for (a)

		civic values	civic values	schools		conflict resol, (a) Coexistence
Measures to reduce domestic violence and child abuse	Outreach programs to protect children at risk of domestic violence	Projects to improve child care, reduce risk factors and promote parental supervision	Various programs now implemented via Instituto Col. de Bienestar Familiar (ICBF)	No assessments available	(a) The IDB/MSSP component Family Educ. project
	Red de Buen Trato al Menor	Network for Prevention of Child Abuse	Active in comuna 20, Partic of Mpal Adm, & NGOs	Incipient; no assessment available	
	Strengthening of the Family Commissaries	Legal advice, psychological counseling, interpersonal conflict resolution, basic family therapy	7 Family Commissaries currently exist in Cali	No assessment available, but the IDB/MSSP foresees a participative diagnostic	IDB/MSSP components: (a) streng. of Family Commissaries; (b) creation of mobile Family Commissariat
Police and community	Upgrading the Central Community call center	Center handles No. 112 emergency calls	Operative; has improved performance over last years	Lack of resources	IDB/MSSP component mod of comm. systems
	Closed Television Circuits (CCTV)	Integrated camera surveillance of hot spots	Nine cameras are operative	Lack of resources	Part of IDB/MSSP component of police mod
	Focused interventions in hot spots in city center	Recuperation of physically degraded sectors	Focused operations undertaken	No assessment available of impact	
	Escuela de Policia Civica Juvenil	Training for neighb youngsters on role of police, human rights, tolerance;	Exist in each urban police Station,	No external assessment available;	An IDB/MSSP component to improve and further develop;
	Escuela de Seguridad Ciudadana	training of adults as community guards	supposedly operates in every police station	no external assessment available	an IDB/MSSP comp to develop selection criteria for candidates
	Frentes de Seguridad Local.	Block watch. Operates around collectively owned burglar alarm	Booming new phenomenon. In Cali, since 1996	(a) No assessment available, (b) Tension between police and Mpal Adm.	Vecinos y Amigos program
	Comites de Convivencia	Promoting neighb control over violence and crime	Apparently nonexistent	Confusion with Veeduria Ciudadana	The IDB/MSSP foresees creation in pre-selected neighborhoods
Improving access to justice	Casas de Justicia (Houses of Justice).	Strengthen inter-inst. coord., community participation	2 Houses of Justice in Cali	Problems of financial sustainability	The IDB/MSSP foresees assess. to define further strat.
	Arbitration and Conciliation Centers	Conflict resolution services	Created by several institutions.	No assessment available.	
Youth and violence	Outreach to street gangs	Negotiate neighborhood truces with pandillas	Low profile initiatives by parishes, Houses of Justice, NGOs	Lack of experience in Cali with local peace negotiations	
Outreach to youth at risk of getting involved in violence	Employment		Successful one time experience with employment of youth in public works	Lack of identification, referral and follow-up of youngsters at risk	2 IDB/MSSP components: (a) Socializacion y Trabajo para Jóvenes, (b) Proyectos de

					Iniciativas Juveniles
	Crime offenders		Smaller projects, mainly undertaken by NGOs and religious institutions	Lack of programs from recently created Vice Ministry of Youth	An IDB/MSSP project Menor Infractor y en Riesgo, will develop method. for intervention
	Promotion of sport	Offer sport facilities to promote alternative youth socialization	Secretary of Sports, with neighb orgs, installed sport fields	Functionality depends on subsequent administration by nghbrhd org. No assessment available	Ongoing
	Ciclovias, free concerts, etc.	Closing of main city arteries during holidays for sports. Organization of free concerts	The Mpal Adm supported by private sector in these activities	No assessment available	Ongoing
	Youth Houses	Aim at risk youngsters of age 12-24	Some have been created	No systematic analysis is available	National initiative
Penitentiary institutions	Reorganization of the Juvenile Rehabilitation Center Valle del Lili	Strat prop for long term solution of juvenile justice adm.	Working group on penitentiary institutions is inoperative	Local adm. has not developed long term strat.	
Attention for internally displaced persons	Integral approach to internal refugees critical issues	Public and private inst intervene on an ad hoc basis	There is no integral municipal program	(a) Internal refugees difficult to reach, (b) Mistrust of refugees	

Source: Martin, 2000.

Strategic priorities for violence reduction in Cali

Long-term strategies for violence reduction and promotion of peaceful coexistence, in the context of the Cali CDS, should consider the following issues: (a) violence in Cali arises from a complex mix of causes, that require a coherent, integral, multi-sector approach and the participation of public and private actors; (b) strategic priorities should build on the city's vast experience in violence reduction; (c) there is now a critical mass of popular support to continue efforts to decrease violence; (d) implementation of violence reduction strategies may provoke violent reactions from armed groups, and it might be difficult to carry out comprehensive community initiatives due to the existence of micro-territories dominated by violent groups, and to conflicts among neighborhood organizations; (e) effective interventions depend on a close reading of the complex political power structure of neighborhoods, which may prove difficult even for community workers; and (f) Cali's efforts to reduce violence and promote citizens' security depend on national policies to address the root causes of the current crisis.

With these ideas in mind, the members of the thematic group on violence and peaceful coexistence propose to integrate the following five strategic priorities in a City Development Strategy for Cali:

- *Improve inter-institutional coordination*. The problems are: (a) deficient inter-institutional coordination and lack of municipal priority given to public security; (b) a risk that projects, committees, and councils will proliferate without effective coordination, leading to diffusion of responsibilities and public confusion; (c) deficient monitoring and evaluation of past and current violence reduction policies; (d) insufficient knowledge of relevant experiences in other cities, such as Medellin and Bogotá, and of relevant international experiences with best practices; (e) limited institutional memory due to high rotation of functionaries.

Desired improvements: (a) strong coordination and management of programs to reduce violence and promote peaceful coexistence; (b) permanent inter-institutional cooperation at all administrative levels, and strong public-private partnerships; (c) formulation of a yearly *Plan de Seguridad Local*, as proposed by the IDB/MSSP program, to include presentation and evaluation of crime data for Cali, with particular attention for its most violent neighborhoods; and assessment of all relevant public and private initiatives underway; (d) strong participation and commitment by all relevant public and private institutions.

Proposed actions or instruments: (a) the IDB/MSSP program should be given high prioirty; (b) an external evaluation of the currently violence reduction policies programs, to better define the roles of DESEPAZ, *Unidad Coordinadora Municipal, Comite de Seguridad y Convivencia, Veedurias Ciudadanas*, and other actors; (c) create knowledge and define strategies on the important issues (see below) of juvenile delinquency, serious juvenile offenders, internal refugees, and street gangs; (d) define precise quantitative indicators for violence reduction in Cali.

- *Recovery of public space.* The problems are: Public spaces have lost their traditional function of promotiong socialization and peaceful coexistence because they have been invaded by the informal economy, gangs, delinquents, drug consumption, and drug dealing activities. Neighbors and others are scared away because of insecurity and physical degradation of the areas. Until now, interventions such as the program *Cali Centro* have not received sufficient priority, and have not been based on an integral multi-sector strategy.

Desired improvements: (a) Physical upgrading of public spaces in cooperation with neighborhoods, so at to promote public appropriation of these places; (b) continuous and appropriate forms of law enforcement to prevent crime and violence from returning.

Proposed actions or instruments: A program to promote community appropriation of public spaces, including: (a) selection of areas for intervention; (b) diagnosis of relevant problems and of neighborhood perceptions of public space and insecurity in these areas; (c) assessment of already existing and related programs and new initiatives. Finally, the thematic group on violence and peaceful coexistence propose a workshop in Cali to share experiences from other Colombian cities and abroad.

- *Improve trust between police and community.* The problems are: While the metropolitan police have made significant progress in efforts to reach out to the community, mistrust is still widespread, and possibilities for the public to express their needs and priorities are still relatively limited. The distance between the police and communities are still far too wide.

Desired result: Institutionalized channels for communication and cooperation between the police and communities.

Proposed actions or instruments: (a) an external analysis of relations and mutual perceptions between the police and communities in Cali in general, and in the most violent neighborhoods in particular, to include proposals for structural improvement[57]; (b) organize workshops in Cali to exchange experiences on community policing, based on national and international experience; (c) design and implementation, by an existing agency, of a community policing strategy for Cali that is consistent with the existing National Program for Community Policing; (d) design and implementation of mechanisms to address corruption within the police.

- *Deal with impunity of serious juvenile offenders.* The problems are: Impunity of serious juvenile offenders, as a consequence of (a) non-existence of an appropriate high-security juvenile correctional facility; and (b) operational deficiencies of the Juvenile Rehabilitation Center Valle de Lilli.

Desired results: (a) safe and secure seclusion of serious juvenile offenders in an appropriate correctional facility; (b) modern and professional seclusion facilities at Juvenile Rehabilitation Center Valle de Lilli for first-time offenders and other non-serious juvenile offenders; (c) creation of a second Juvenile Rehabilitation Center in the northern part of Valle de Cauca; (d) as a result of (a) through (c), improve trust between the justice administration and the police, and diminish popular support for practices of social cleansing against (supposed) juvenile delinquents.

Proposed instruments: (a) activate the inoperative *Comite del Menor Infractor* so as to systematically assess juvenile delinquency issues and make them a priority; (b) hire an external expert to assess the situation of serious juvenile offenders in Cali in terms of impunity, escapes, rehabilitation, etc., and develop long-term policies to address these issues; (c) hire an external expert to assess Valle de Lilli in terms of internal organization, escapes and other security problems, rehabilitation programs, etc., and develop long-term policies; (d) assess, in coordination with the Ministry of Justice and other relevant national and sectional agencies, possibilities for construction of a regional High Security Juvenile Correctional Facility for serious juvenile offenders, and develop short-term alternative solutions; (e) learn from foreign experiences regarding correctional facilities for serious juvenile offenders.

- *Attention to youth and street gangs.* The problems are: Youth and street gangs are systematically identified as a major problem in the most violent neighborhoods of Cali, but there are few policy initiatives designed to help them.

Desired result: A coherent outreach program directed toward youth at risk of entering street gangs. Such a program should be part of an integral policy to reduce juvenile delinquency through education, sports, cultural programs, employment programs, and efforts to increase social capital.

Proposed actions or instruments: (a) assess the character of youth who participate in street gangs[58]; (b) workshops to share experiences from the local, national, and international levels; (c) studies to understand how "wise men and women" and "big brothers and sisters" can help to promote reconciliation[59]; (d) creation of a public-private Social Fund, in line with *Plan*

---

[57] In 1999, Rodrigo Valencia de la Roche, coordinator of the *Casa de Justicia* in Aguablanca, presented to the IDB/MSSP program a proposal for a *Proyecto Consulta Ciudadana* on this issue. The proposal was inspired in part by a similar effort in Bogotá, in 1996, by the *Centro Nacional de Consultoría*.

[58] The South Bank University in London intends to cooperate with the *Fundacion para la Asesoria y Progreso de la Salud* (FUNDAPS) and the Secretary of Health, to assist youth groups with carrying out a social capital study that will focus on relations between youth and violence in eight of Cali's most violent *comunas*.

[59] Compare with the success of the Women Creators of Peace project in *barrio* Cazuco in Bogotá (Klevens, 1998, p.22).

Nacional para el Montaje de Escenarios de Convivencia, to implement an outreach program for youth at risk of joining gangs.

The thematic group identified four additional issues that need to be addressed in coordination with other thematic groups involved in the Cali CDS, and one that depends far more on national than on local policies. These are: (a) a major educational effort, with particular attention to increasing attendance, decreasing absenteeism and dropping out, improving educational quality, and promoting civic values; (b) particular attention to high levels of youth unemployment; (c) promotion of good governance to set a good example for society and increase the credibility of public institutions; (d) promotion of a coordinated outreach program targeting internal refugees; and (e) strengthening of national law enforcement.

6. EDUCATION

Introduction

It is widely accepted among social scientists that the dynamics of human and social capital formation are critical to explaining the poverty and extreme social inequalities typical of Colombia and other Latin American countries. Likewise, human and social capital investment is considered a key factor in sustainable development; and education, in particular, is a necessary component of any strategy for social change. In Cali, several problems have been identified in the education system: (a) it is not closing the gap between social classes; (b) it is not providing young people with the skills to participate successfully in the labor market; and (c) its quality and coverage have worsened over the years. These deficiencies are considered by members of the private sector, the universities, the government, and civil society organizations to be one of the worst obstacles the city is facing.

This chapter reflects debates and analyses of education resulting from the City Development Strategy process. The chapter has three subsequent sections. The first section describes the current education context in Colombia, with its present achievements and challenges. This section is relatively long and goes into some detail, because the education situation in Cali is strongly linked to the situation nationwide. The second part focuses on education in Cali. The third part proposes short-term strategies for alleviating the effects of the crisis in education, and broad strategic areas for developing education in the city in the coming years. It is important to note that: (a) tertiary education is not explicitly addressed; and (b) information on education in Cali is highly deficient, in terms of both quality and quantity.

Education in Colombia

More than 8.5 million children attend 57,000 public and private schools in Colombia. The contribution of the private sector to education is higher in Colombia than in most Latin American countries: about 15 percent of all primary schools are private, and about 40 percent of secondary schools. This private support to education represents about 3.6 percent of the GNP. As for the public sector, during the 1990s, government investment in education increased substantially. In 1990 the sector received amounts equal to 2.5 percent of GNP, and this rose to 4.1 percent in 1997. The participation of education in the national budget is more than 20 percent. Public schools are supported principally via national transfers *(transferencias)* from the central government to the provincial and municipal governments. Local governments participate with additional but relatively small amounts.

Education in Colombia has undergone many changes in the last 15 years. This section illustrates some of these changes, first stressing the achievements, then pointing out the most persistent problems. As in the rest of Latin American and Caribbean countries, access to school for all children has substantially improved over the last two decades. Primary school net enrollment rates increased from 62 percent in 1935 to 83 percent in 1997. In 1985, only three out of ten Colombian children were enrolled in secondary education. Today more than six out of ten have this opportunity (please refer to Graph 6.1).[60]

Other important <u>achievements of education in Colombia</u> include:

- High participation of females in all levels of education. In 1992, for example, the mean years of schooling for people over 25 years old was higher for females than for males (7.7. and 7.3, respectively).[61]

---

[60] Data of the Departamento Nacional de Planeación (see http:www.dnp.gov.co/sisd)
[61] UNDP, 1994.

- The development of evaluation mechanisms to improve the quality of education. Colombia has one of the oldest testing services in the hemisphere.
- The implementation of world-class educational innovations. *La Escuela Nueva,* for example, has inspired several rural education models in other countries; and research conducted in early childhood education in Colombia has contributed to the development of preschool education throughout the Latin American region.
- Civil society organizations (popular organizations, foundations, community-based institutions, etc.) have played a critical role in the development of formal and non-formal education in the country. More than 40 percent of the 1,371 civil society organizations analyzed in 1992 by the *Fundación Social* were involved in educational activities.[62] (Vargas et. al, 1992).

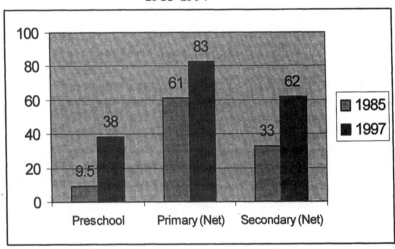

Graph 6.1.
School enrollment rates in Colombia
1985-1997

Source: Departamento Nacional de Planeación, 1999.

The education sector in Colombia has limitations of supply, quality, and institutional management. In spite of the improvement in primary and secondary education coverage (Graph 6.1), limitations persist: 17 percent of Colombian children aged 7-11 do not attend primary school, and 38 percent of those aged 12-17 do not attend high school. Moreover, these deficiencies affect mostly the poor, who, in Colombian cities, have an average of 6 years of school, while the non-poor have 10.3 years.[63] Graph 6.2 shows the percentage of poor, middle, and richest youth (ages 15-19) attaining different levels of education in Colombia. Poor youngsters exhibit significantly lower educational attainment in all grades than their wealthier counterparts, and this difference grows with years of schooling.

Moreover, despite all reforms and strategies implemented over the last decade, the quality of schooling continues to be low in terms of content, as measured by criterion reference tests and in comparison with international standards. National evaluations conducted by the Ministry of Education show deficiencies in basic reading and mathematics in grades 3, 5, 7, and 9.[64] Colombia, the only Latin American country participating in the Third International Study of Science and Mathematics (TIMSS), ranked 42 among 43 countries in 1996. Within Latin America, Colombian children do not perform above the median of their

---

[62] Vargas et al, 1992.
[63] World Bank, 1999.
[64] *Ministerio de Educación,* 1997.

counterparts in neighboring countries. This is illustrated in a recently published UNESCO study,[65] which compares the language achievement of students of several Latin American and Caribbean countries (please refer to Graph 6.3).

Graph 6.2. Educational attainment in Colombia, 1995
(highest grade attained by youth ages 15-19)

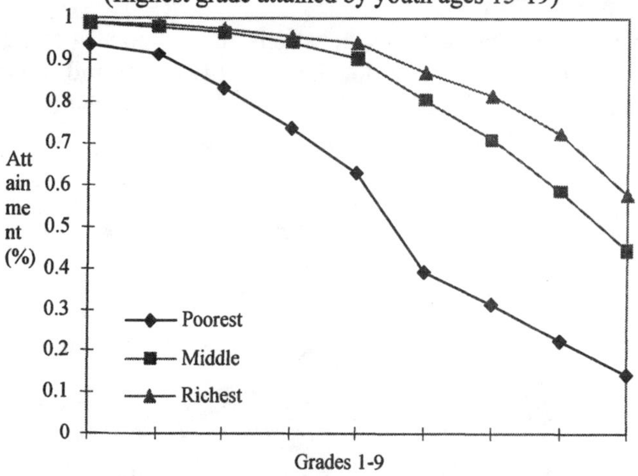

Grades 1-9

Source: Filmer, and Pritchett, 1999.

Perhaps the most critical problem in Colombia's education sector is poor management. An incomplete decentralization process, initiated several years ago, left incoherent decisionmaking processes, weak and dependent schools, a lack of incentives for teachers and administrators, and few accountability mechanisms.

Education in Cali

Cali's education indicators do not differ significantly from those of other large Colombian cities. Children ages 7-11 have higher rates of participation in the schools of Bogotá and Bucaramanga than in Cali; and children ages 12-17 have a higher probability of being in school in Bogotá and Barranquilla than in Cali (please refer to Graph 6.4).

In terms of quality, the available indicator, the national standardized test, given to all students in their last year of high school, suggests that almost three quarters of the secondary schools in Cali (74 percent) perform at medium and low levels. Only 6 percent of the schools can be classified as very superior. This deficient performance is not very different from that of schools in Medellín (please refer to Table 6.1).

The impact on education of the general crisis in the country is noticeable in Cali. The many problems in Cali's schools are presented under three headings: decline in school attendance; limitations in the provision of education, in terms of both quality and quantity; and institutional obstacles. All of these problems are making the persistent socioeconomic inequalities in Cali even worse.

---

[65] UNESCO, 1999.

Graph 6.3. Third grade language achievement scores in Latin America, 1999

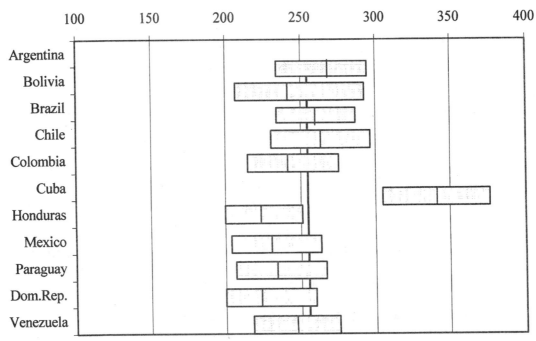

Source: UNESCO, 1999.

Graph 6.4. School participation rates of children 7-11 and 12-17 in five Colombian cities, 1997

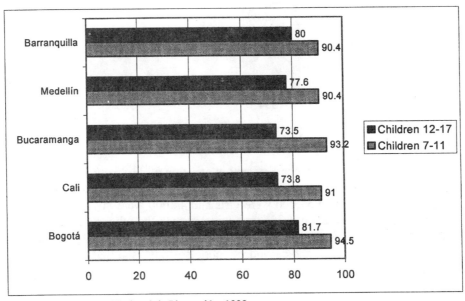

Source: Departamento Nacional de Planeación, 1999.

Table 6.1. School performance in Cali and Medellin, 1999
(percentage of schools)

	Medellín	Cali
Very superior	4.0	6.0
Superior	5.0	6.4
High	15.0	13.6
Medium	20.0	29.0
Low	51.0	36.0
Very low	5.0	9.0

Source: Servicio Nacional de Pruebas, 1999.

School attendance in Cali has worsened in recent years. After sustained and successful efforts to enroll all children in primary education throughout the country, rates of attendance for children 6-12 years old in Cali have fallen from 98.26 percent in 1994 to 93.50 percent in 1998, according to the results of national household surveys (please refer to Graph 6.5).[66] The most recent information, collected by the EPSOC survey (please refer to Table 6.2), shows an overall primary school attendance rate of 94.8 percent in 1999—a 3.46 percent decrease since 1994. This means that about 15,000 youngsters are not in school, two thirds of them boys. Another characteristic of primary school attendance is the gap between the lowest and highest income quintiles: 91.1 and 100 percent, respectively.

Graph 6.5. Attendance in Cali primary schools, children ages 6-12, 1994-1998

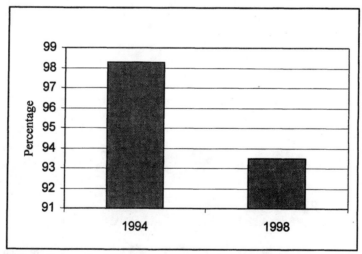

Source: Santamaria, 1999.

Older children (ages 13-18) attended secondary school at a rate of 73.8 percent in 1997 and 70.1 percent in 1999[67] —a significant 3.7 percent decrease. This means that about 100,000 youngsters were not enrolled in school in 1999, of which 57,000 were girls and 44,000 were boys. The richest quintile of the population shows stronger participation in higher education than the other quintile groups (85 percent in the highest income quintile versus 58.4 percent in the lowest). The EPSOC survey found that the main reason for low enrollment in primary and secondary schools is cost, and that this effect is more severe in the poorest income group.

---

[66] Santamaría, 1999 (a)

[67] *Departamento Nacional de Planeación,* 1997, and EPSOC 1999, respectively.

Table 6.2. Education and poverty in Cali, 1999

School attendance	Income quintile					Total
	1	2	3	4	5	
6-11 years	91.1	93.8	95.3	97.9	100	94.8
- private	32.6	31.9	39.7	62.3	72.2	100.0
- public	67.7	69.0	58.5	36.3	25.0	55.7
- male	86.9	95.1	92.8	100.0	100.0	93.9
- female	95.4	93.6	97.0	96.7	100.0	96.1
12-18 years	58.4	68.9	69.9	73.1	85.0	70.1
- private	43.1	50.1	54.8	56.6	74.9	100.0
- public	56.4	50.9	44.1	42.0	22.3	43.2
- male	60.5	72.8	67.1	69.1	87.9	71.0
- female	57.0	65.6	72.7	76.0	82.1	69.3
Reasons for not attending						
- costs	59.5	57.1	43.7	36.7	13.4	48.8
- work	2.1	7.2	9.9	18.8	17.2	8.6

Source: EPSOC, 1999; Hentschel, Mehra, and Seshagiri, 1999.

Where do children go when they leave school? Some find a place in the work force, while others stay in the streets and contribute, as either perpetrators or victims (or both), to violent acts in the city. In Cali, the number of children working inside (taking care of younger children, cooking, doing laundry) and outside the home has increased in recent years. The proportion of children 12-16 years old contributing to the family income by working outside the home rose from 7.9 to 9.1 percent between 1994 and 1998.. During the same period, the proportion of children working at home to allow adults to enter the labor force increased from 9.6 to 14.5 percent.[68]

Children and youngsters who neither attend school nor are engaged in work are at high risk of getting involved in illegal activities. A high percentage of the victims and perpetrators of homicides in Cali are young (please refer to Chapter 5). In addition street gangs involved in drug dealing, prostitution, and robbery are proliferating in the city and exhibit extremely violent behavior.

The <u>quantity and quality of the education supply</u> have worsened in the last years. Municipal investment in education in Cali is significantly lower than the level of investment of other large cities in Colombia (8.4 percent of the total municipal budget in Cali, versus 57.3 percent in Medellín, 23.8 percent in Bogotá, 39.4 percent in Pereira, and 16.1 percent in Bucaramanga). This limited investment of resources manifests itself in the deterioration of Cali's public schools. The performance of public school students, measured by learning outcomes, indicates that public schools are not improving in comparison with private schools. In the 1996-1997 school year, more than half of all public schools did not perform up to the standards established by the National Testing Service, and only 10 percent excelled in achievement, while 26 percent of private schools were high achievers.[69] Moreover, institutional reforms providing more authority to public schools have not had the expected results[70] in terms of efficiency and effectiveness of the system.

Of the nearly 400,000 students enrolled in the school system in Cali, only 157,000 are enrolled in public schools. The other 243,000 are served by private schools. However, the crisis in Cali is having a strong impact on private education as well. More than 45 private schools in the Valle del Cauca Department—the majority of them in Cali—have closed because students are not enrolling in the same numbers as in

---

[68] Santamaria, 1999. Data from household surveys for 1994 and 1998 confirm this finding
[69] Alcaldia de Santiago de Cali, 1998.
[70] Fundación Foro Nacional por Colombia (1998) Estudio sobre el gobierno escolar.

previous years.[71] This has led the Association of Private Schools to call for the declaration of an educational emergency. At the same time, low and middle-class families are abandoning private schools and enrolling their children in less expensive public schools, thus adding to the deficit of space in public classrooms.

One manifestation of the deteriorating quality of education in Cali is the high unemployment rate for high school graduates. Unemployment in Cali particularly affects young people with a complete secondary education. While 19.1 percent of the work force without education is unemployed, those who have completed secondary school have a 25.1 percent unemployment rate, while those with a higher education have a 7.9 percent unemployment rate (please see Table 6.3).

Thus it is clear that secondary education is not providing young people with the skills to successfully enter the world of work. Such problems are caused not only by the economic recession, but also by the failure of schools to adapt their curricula to the changing knowledge and skill needs of productive enterprises, in the context of the new global economic environment.

Table 6.3. Unemployment levels by educational attainment in Cali, March 1999

Educational attainment	Unemployment rate
Without education	19.1
Incomplete primary	19.2
Complete primary	17.8
Incomplete secondary	24.2
Complete secondary	25.1
Incomplete superior	27.1
Complete superior	7.9
Total	21.2

Source: Urrea and Ortiz, 1999, p. 38.

<u>Private returns to education</u>. Estimated private returns to education support the conclusion that secondary schooling in Cali has comparatively low returns. While primary schooling has—on international levels—a comparatively good return (please refer to Table 6.4), with every year of primary schooling increasing expected earnings by about 9 percent, returns to secondary schooling are considerably lower (at 6.5 percent). Returns also differ with age. For the working population below 35 years of age, all returns are several percentage points less.[72]

Table 6.4. Yearly private rates of return to education in Cali, September 1999

Years of education	Total work force	Work force < 35 years
Primary school (years 1-5)	9.3	6.0
Secondary school (years 6-11)	6.5	5.5
Superior (years 12 and higher)	7.5	6.4

Note: The yearly rates of return for the work force below 35 years of age are not significantly different from each other.
Source: Hentschel, Mehra, and Seshagiri, 1999.

---

[71] Estimated data of El País, 1999.
[72] These estimates are calculated only for the occupied population and do not take the unemployed into account. Hence, average returns over the whole labor force (employed and unemployed) are lower.

Finally, there are several <u>institutional obstacles</u> to improving the education system in Cali: (a) there are no information or monitoring mechanisms by which policymakers, communities, and families can evaluate the system's performance; (b) political clientelism is an obstacle to institutional change; (c) schools lack autonomy, and there are conflicts beween provincial and municipal-level management of the system; and (d) civil society organizations involved in educational activities face increasing economic difficulties.

Proposed strategic priorities for education in Cali

A total of five strategic priorities are proposed and explained in this section.

- *Protecting children from dropping out of school*. Children should be protected so they can stay at school. The impact of thousands of children in the streets or in the labor force will be devastating to their personal and social development as well as to the social fabric of the community. Since we do not have a clear understanding of the critical factors involved in schooling decisions, rapid surveys may provide additional information so that preventive programs can be developed. Research centers will have an excellent opportunity to contribute to policy making in this area.

Emergency measures may include the provision of incentives for families to keep children in school. Some of these incentives could take the form of complementary services for children at school, such as nutritional supplements and health care. Assistance to families and communities to create better learning environments for children at home and in the neighborhoods, and to demand better services and higher outcomes from the educational system, would be particularly useful. Other incentives may be financial, such as the provision of textbooks, supplies and uniforms, offering scholarships, or a combination of loans and scholarships. Student loans can be claimed by school attendance and achievement and special contributions to society.

- *Support to youth development programs.* Youth are in the middle of the crisis. It is essential to provide them with opportunities for learning, working and social commitment. The government, the private sector and the organizations of the civil society could work together developing a youth strategy for the city. There is already a history of good youth programs in the city of Cali that should be considered. The following recommendations could contribute to this process.

    - A set of criteria should be developed to be used in the assessment of youth programs. The most effective youth programs in Cali should be identified, supported and adapted to other city contexts.

    - Young people could also get involved in the development of the city via special projects supported by governmental and non-governmental sources. They can obtain college scholarships, a small salary, and training opportunities. Young leaders in business, politics, or sciences could be identified and helped.

    - Young people, particularly women, could help reduce the deficit of preschool education in the city. Short training courses would prepare them for implementing alternative preschool programs in all *comunas* of Cali, and particularly in the poorest ones.

    - Schools could offer after-school programs for youth and offer complementary services or connect adolescents with existing services for them in the city. *Proposed actions or*

This recommendation coincides with one of the recommendations in Chapter 5, "Attention to youth and street gangs". Both should be consolidated in one comprehensive approach.

- *Creation of a program to improve the contribution of education to the city's labor supply.* This program would work toward a better coordination of formal and non-formal education to facilitate an easy school to work transition for young people, and the preparation of the human resources that the productive sector requires. The program would include:

    - The reform of secondary schools. Secondary schools should provide youngsters with the competencies required for entering into the labor market. International standards in mathematics and science should guide curriculum development. Special attention should be given to the informational technology. Students and parents should have voice in the management of the system.

    - The coordination of non-formal education, youth programs, and training so students may have diverse and holistic learning opportunities.

    - Opportunities for all secondary students to have hands-on experience in the use of advanced technologies (This program could be supported by private companies).

    - Internships providing work experience in private and public organizations.

- *Promote the inclusion of civic values in education.* The purpose of this strategic priority is to improve social cohesion, the participation of young people in the democratic process, and the strengthening of the value system supporting stable peace and democracy in the country. This strategy may involve the support of formal and non-formal education through teacher training on managing "democratic schools and classrooms", and the formulation of learning standards related to "solidarity skills", conflict resolution and civic participation that all students should exhibit at the end of high school. These learning outcomes should be incorporated in the national tests. Illustrative activities of this strategy are:

    - Provide scholarships and other incentives for youth to participate in activities promoting civic values and solidarity.

    - Involve mass media in the creation of interesting programs discussing issues of social and civic interest (an example is the television series "Grado Doce" developed by the Social Foundation.

- *Creation of an education management mechanism.* Education is one of the largest and most critical enterprises of the city of Cali. It requires a small, flexible and learning oriented management system that could be implemented by an experienced and respected NGO, and with the participation of relevant stakeholders. Its main roles would be proposing municipal educational policies, providing reliable information, such as school demography and financial statistics[73], and assessing school and teacher performance and the output of the system. Private and public schools need guidance, monitoring and support to improve the quality of their services.

---

[73] The development of the Municipal Educational Plan 1998-2,000 found this information to be unreliable.

Main components of this mechanism could be:

- A continuous evaluation operation that would develop standards for students and schools and report to citizens and government. Such a system would replace current ineffective supervision and inspection units.

- A fund for special studies and analyses to be granted to research centers, universities and private organizations on competitive basis.

- A monitoring and communication unit providing information to individual schools, students, families and citizens on the outcomes of schooling, and processing feedback to the system.

- The creation of incentives for the development of the learning industry in the city (publishing business, educational software, educational television, etc.), and the participation of churches, sports clubs, private industries and mass media in education.

7. FINANCIAL SITUATION

Introduction

At the beginning of 1998, Cali's new municipal administration faced a difficult situation. The city's economy was in recession and the municipal budget was unbalanced, due to excessive borrowing and a disproportionate number of employees. This situation worsened in the second half of 1998, when Colombia's economy slowed even further, as a result of a drop in commodity prices, a drying up of capital inflows, and the tightening of domestic monetary and fiscal policies to defend the peso exchange rate band. When the Colombian interest rate moved up in 1998, the financial expenses of the Cali municipality increased by 80 percent in relation to 1997, reaching more than 200 percent of the gross current balance. As a consequence, the municipality did not service its debt from August 1998 until May 1999.

The next section of this chapter analyzes the municipality's historical financial data up to September 1999, when information for this report was collected. The subsequent section reviews the financial position of the municipal companies. The fourth section considers the financial perspectives of the municipality for the period 1999-2003, including and excluding the Metro project. The final section summarizes the main financial problems the municipality is facing, and presents the strategic priorities discussed with the local counterpart team in September 1999, as part of the CDS exercise. It is important to note that the analysis found inconsistencies among the different sources of financial information, and an incorrect accounting of certain items under disguised figures.

The central administration

The services provided by the municipal administration are urban planning and housing, health and education, social assistance, transport, thoroughfares, public safety, economic promotion, environment, culture, and sports. <u>Expenditures</u> are concentrated in policy formulation, administration and finance (41.6 percent), health and education (21.5 percent), and transport and thoroughfares (17.6 percent). Social assistance expenditures, including disaster prevention and mitigation, represent a very low 1.4 percent. At the bottom of the list, economic promotion accounts for only 0.4 percent (please refer to Table 7.1).

When expenses are broken down by type rather than by activity (please refer to Graph 7.1), current personnel expenses represent 30 percent, excluding debt service. In fact, additional personnel expenses are erroneously classified under investments (13 percent). Total personnel in the central administration was 9,174 in 1999, of which the majority, 7,774 are accounted for as current personnel expenditures and an estimated 1,200 to 1,400 are hidden under investments. Further inquiries revealed that the latter are temporary staff, with high salaries, appointed mostly through political influence and considered highly strategic. Total personnel expenses thus represent 43 percent of total expenses and more than 80 percent of current expenses (total expenses minus investments). This figure is extremely high compared to other municipalities worldwide, where expenses range from 20 percent to 40 percent.

From 1991 to 1997, the staff did not increase, except in the City Council and in the Health and Education units, the latter due to the transfer of these services from the national to the local governments. These employees work under various types of contracts (please refer to Table 7.2). Public contracts are administrative contracts for civil servants appointed by a public authority. Official contracts are temporary labor contracts for personnel who execute or maintain public works. In practice, of the 1,575 employees classified under official contracts, 842 should be under public contracts, as they are cleaners, messengers, drivers, etc. The payroll is prepared by the Human Resources Unit, based on information provided monthly by each unit. It is estimated that 90 percent of the staff get their jobs thorugh political influence, and only 10 percent based on professional capacity and motivation. Municipal staff, in general, earn higher salaries than they could in the private sector.

Table 7.1. Breakdown of expenditures by activity, 1998 (excluding debt service)

Activity/Program	Total expenditures US$ million[1]	Percentage of municipal budget
Admin., finance, and political direction	89.78	41.6
Health and education	46.56	21.5
Transport and thoroughfares	37.89	17.6
Environment	12.58	5.8
Public services and others	6.91	3.2
Urban planning and housing	5.99	2.8
Culture and sports	4.98	2.3
Public safety	4.94	2.3
Social assistance	2.98	1.4
Institutional modernization	2.46	1.1
Economic promotion	0.79	0.4
Total	215.86	100.0

Note: Figures used in the Financial Adjustment Plan show total expenditures of U$181.6 million.
Source: Municipio de Santiago de Cali, 1999.

Graph 7.1. Breakdown of total expenditures by type, 1998
(percent, excluding debt service)

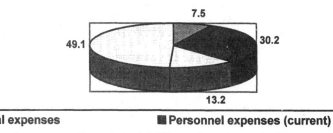

| ■ General expenses | ■ Personnel expenses (current) |
| □ Personnel expenses (investments) | □ Investments, excluding personnel |

Source: Municipio de Santiago de Cali, 1999.

Although the Centers of Integrated Local Administration (CALIs) manage a small amount of the municipal budget, most of their staff are concentrated in central administration headquarters. In terms of pensions, in 1999, the number of pensioners was 3,515, with an annual cost of US$18 million. The Colombian local governments might capitalize the pension plan, but, in general, they are paying the pensions each year, charging the costs to the annual budget.

In terms of current revenues, own-source revenues—locally raised revenues plus fines and other revenues—represent 67 percent of current revenues. National government transfers represent 31.6 percent, and the remaining 1.4 percent come from contributions to the pension plan made by active and nonworking –personnel (please refer to Table 7.3). National government transfers are conditional funds that can only be used for specific purposes, such as health, education, housing, sports, and planning. As a consequence, the city's own-source revenues have to finance all remaining municipal services. Real estate taxes and industry and commerce taxes represent the single largest sources of revenues.[74]

Table 7.2. Breakdown of number of employees, 1999

Type of contract	Number of employees	Average annual salary (US $)
Public contracts	5,016	8,110
Official contracts	1,575	6,577
Council	251	10,670
Control entities	761	11,886
Service contracts	1,400	n/a
Total	9,174	

Source: Unidad de Recursos Humanos, Municipio de Santiago de Cali, 1999.

The City Council has to approve a minimum and a maximum tax rate every year. Tax rates in Cali are in the upper level and have been increasing rapidly: from 1991 to 1997, the Cali consumption price index has multiplied by 3.1 and the taxes collected by 7.4. These increases, particularly when not accompanied by an improvement in municipal services, promote tax evasion. In the case of Cali, citizens think the rates are high, the quality of services is poor, and services are better in areas of the city with higher incomes. In addition, since 1993 total real estate tax collection has been greater than total industry and commerce tax collection (please refer to Graph 7.2), meaning that the fiscal pressure falls more on families than on businesses.

Table 7.3. Breakdown of current revenues, 1998

	Current revenues (US$ million)	Percent of municipal revenues
Real estate taxes	46.83	24.0
Industry and commerce tax	34.85	17.9
Gas surcharge	9.49	4.9
Other taxes	8.03	4.1
Local taxes	99.20	50.9
Fees and contributions	17.04	8.7
Locally raised revenues	116.24	59.6
Fines and other revenues	14.50	7.4
Transfers	61.55	31.6
Pension plan	2.61	1.4
Total	194.9	100.0

Note: Figures used in the Financial Adjustment Plan show current revenues of US$137.7million.
Source: Municipio de Santiago de Cali, 1999.

---

[74] The principal taxes of the Colombian municipalities are:
- Real estate tax, charged on the cadastral value of properties at a rate of 2 to 16 percent
- Industry and commerce tax, levied on previous year gross revenues
    * between 2 and 7 percent on industrial activities
    * between 2 and 10 percento on commercial and service activities
    * 3 percent on savings and loans associations
    * 5 percent on other financial intermediaries
- Gas surcharge, charged on gas price at a rate of 14 to 15 percent.

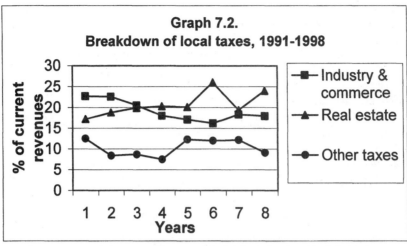

Source: Departamento de Recursos Humanos, Municipio de Santiago de Cali, 1999.

Investments in Cali have multiplied by 9.6 from 1991 to 1997, while current revenues have multiplied by 7.8. As a consequence, the municipality has incurred debt as a way to maintain its investment program, while investments have been reduced over time in response to liquidity problems. According to the Administration and Budget Execution Division, the 1999 investment program was US$166 million (please refer to Table 7.4.). In the first half of 1999, the degree of investment execution was 27 percent, with the largest proportion in the thoroughfare and public transport category. This reflects the high priority that the municipality gave to the Metro project. At that time, no investments had been executed for housing, and very little for education and environment.

Table 7.4. Investment program, 1999

Activity	Investments (US$ million)	Investment breakdown (percent)	Percent executed in 1st half of 1999
Thoroughfares and public transport	72.0	43.4	16.3
Health	29.4	17.8	34.4
Environment	18.6	11.2	9.6
Education	13.8	8.3	8.2
Housing	10.4	6.3	0.0
Other	21.7	13.0	44.8
Total	165.9	100.0	27.0

Source: División de Administración y Ejecución Presupuestal, Municipio de Santiago de Cali, 1999.

The city's official budget statements show a deficit from 1991 to 1998. In 1999, this process accelerated when Colombian interest rates moved up, given that most of the municipal debt has variable interest rates. In December 1998, the City Council approved an unbalanced budget, indicating that the items without financing were US$7.7 million in current expenditures, US$41.2 million in debt service, and US$8.0 million in investments.

When the debt information enters the equation, the financial situation of the municipality is even worse. Before starting this analysis, let us review the Colombian regulatory framework for municipal borrowing. Law 358 of 1997 requires that a local government's capacity for debt payment be measured through two ratios: interest as a percentage of gross current balance, and debt as a percentage of current revenue (please refer to Table 7.5). The local government can raise debt with or without approval of the national government, depending on the level of these ratios. When approval is needed—that is, when the financial situation does not meet preset criteria—the municipality has to agree with the national government on a Financial Adjustment Plan.

Table 7.5. The Colombian regulatory framework: borrowing limits

Interest as percentage of gross current balance	Debt as percentage of current revenues	Requirement of the Financial Adjustment Plan[1]
<40%	<80%	No
>40% and <60%	<80%	Depends on whether debt growth is greater than CPI growth * 40%
>60%	>80%	Yes

[1] In all cases, the debt has to be registered with the Public Credit General Department (part of the Ministry of Finance).
Source: Ministerio de Hacienda y Crédito Público, Dirección General de Apoyo Fiscal, 1998.

By 1998, if accrued and nonpaid interest are included, the central administration debt was US$196 million, consisting of US$139.4 million in capital and US$56.6 million in accrued interest. The accrued interest increased by 80 percent, due to the following factors: (a) a major part of the debt is floating debt; (b) municipal deficits have constantly increased; and (c) the monetary policy applied by the national government to defend the Colombian peso has affected interest rates. As a result, from August 1998 to May 1999, the municipality neither paid capital nor interest.[75]

On December 31, 1998, nonofficial figures[76] showed that the central administration debt represented 142.1 percent of current revenues (if accrued and nonpaid interest are capitalized as debt), and accrued interest represented 203.6 percent of the gross current balance. Given these results, the municipality was required to prepare a Financial Adjustment Plan.

On June 10, 1999, the plan was signed by the municipality and its creditor banks. The plan's policies for the period 1999-2009, were to: (a) increase locally raised revenues at least at the CPI rate; (b) apply US$25.3 million to reduce personnel in 1999 and 2000 by 1,000 each year; (c) decrease service contract expenditures by 20 percent in 1999 and by 30 percent per year from 2000 to 2009; (d) freeze the number of employees after year 2000; (d) reduce general expenditures by 10 percent in 1999 and by 2.6 percent in 2000; and (e) maintain general expenditure increases at less than the CPI rate after year 2000. The main goal of the Financial Adjustment Plan is to reduce the municipal debt by 57 percent in three years. The forecast for the plan's current revenues and expenditures are presented in Graph 7.3.

By 1999 the banks had refinanced all of Cali's bank loans with a 10-year credit, a grace period of 3 years, and interest rate of DTF[77] plus 300 basic points, to be paid one quarter in advance. Additionally, the banks have conceded a credit of US$37.3 million, with the same terms as above. This credit is to finance the accrued and nonpaid interest from August 1998 until May 1999. As a result, the municipality's debt has increased to US$215 million since the plan was signed. Graph 7.4 shows the expected trend of financial expenses under the Financial Adjustment Plan.

---

[75] The new administration (2001-2003) took drastic measures to change the situation, re-negotiating the debt at better terms.

[76] Included in the Financial Adjustment Plan itself, but inconsistent with official figures. Official figures are: debt= 91.3 percent of current revenues; interest= 46.9 percent of gross current balance.

[77] DTF = Dépositos a Término Fijo Domestic interest rate for quarter deposits.

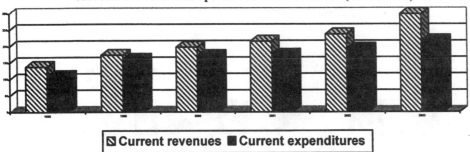

Graph 7.3. Financial Adjustment Plan, 1998-2003: current revenues and expenditures before interest (US$ million)

Source: Alcaldía de Santiago de Cali, 1999 (b).

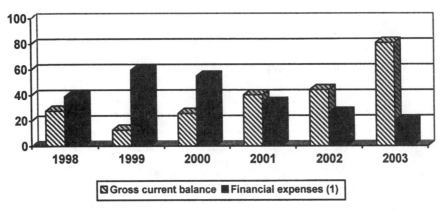

Graph 7.4. Financial Adjustment Plan, 1998-2003: financial expenses[1] (US$ million)

[1] Financial expenses = paid interest.
Source: Alcaldía de Santiago de Cali, 1999 (b).

A second credit of approximately US$22.7 million was scheduled for June 1, 2000 to finance the interest accrued from June 1, 1999 until May 31, 2000. This credit was to have a term of 5 years, a grace period of 2 years, and an interest rate of DTF plus 300 basis points. All bank loans have a revenue collateral equivalent to 150 percent of annual debt service. The plan includes the sale of part of the Cali Municipal Services Company, EMCALI, for US$73.3 million, of which US$10.1 million would pay interest and US$63.2 million would prepay the bank credits. The municipality was supposed to pay such amounts to the banks before June 1, 2000. Additionally, the municipality was required to decrease investments by US$20.2 million in 1999 and by US$75.8 million from 2000 to 2002. Graph 7.5 illustrates the expected trends of capital revenues and investments under the plan.

The municipality agreed to pay US$25.3 million of debt in 2001 and 2002. All of the measures explained above would result in a debt decrease by 57 percent during the period of the plan, as illustrated in Graph 7.6.

**Graph 7.5. Financial Adjustment Plan, 1998-2003:
capital revenues and investments** (US$million)

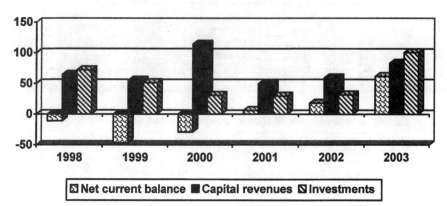

Source: Alcaldía de Santiago de Cali, 1999 (b).

**Graph 7.6. Financial Adjustment Plan, 1998-2003:
debt** (US$ million)

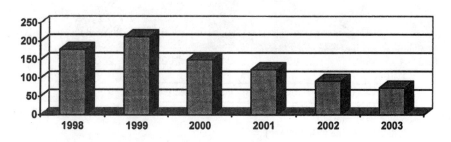

Source: Alcaldía de Santiago de Cali, 1999 (b).

In the year 2000, the critical points of the plan were: (i) the decrease of personnel expenses; and (ii) the sale of a part of EMCALI. In September 1999, the Mayor approved the elimination of all salary benefits that were not regulated by law (extra-legal benefits). This measure was being used as an alternative to reducing staff and has generated public protests and disturbances. As of September 1999, the Cali Council had not yet approved the sale of a part of EMCALI, but agreed that an investment bank be contracted as advisor of the sale. As of December 1999, no such bank had been contracted.

The municipal companies

There are four municipal companies: (a) EMCALI, in charge of distribution of electricity; distribution of water; operation of three plants that purify water taken from the Cali and Cauca rivers; construction, maintenance, and operation of the sewerage system; and telecommunications; (b) EMSIRVA, in charge of collection and disposal of solid waste; (c) CALIASFALTO, an asphalt plant; and (d) BANCALI, in charge of managing the municipal cash balances. Table 7.6 contains the basic financial indicators of these companies. By all accounts, EMCALI is the largest of these companies. The municipal companies supply more than 90 percent of the city's inhabitants with sewerage and electricity, 95 percent with water, and

96.5 percent with solid waste collection. In telecommunications, the share of the market is smaller, because two new private operators are competing in this market.

The prices of the services delivered to residential users are approved by the national and municipal governments, depending on the service provided, production and distribution costs, operating margins, and users' income levels, with users classified according to six strata. Prices are lower than costs for strata 1 to 3 (the poorest), equal to costs for stratum 4, and greater than costs for strata 5 and 6. Each year, the national and municipal governments are supposed to compensate the subsidization process with grants, but these usually remain on paper only. Nevertheless, EMCALI and EMSIRVA view these grants as revenues for accounting purposes.

Table 7.6. Municipal companies, basic indicators

Name	Share capital (percent)	1998 sales / revenues US$ million	1998 operating profits US$ million	1997 number of employees	1997 number of service users
EMCALI	100	116.31	22.42	4,100	Electricity: 419,279 Water: 413,504 Sewer: 392,937 Telecomm: 457,022
EMSIRVA	100	23.49	1.89	653	417,395
CALIASFALTO	100	...	...	4	
BANCALI	100	7.69 (1)	7.59[1]	17	

[1] 1997 figure.
Sources: Empresas Municipales de Cali (EMCALI) 1996, 1997, 1998, 1999; Empresa de Servicio Público de Aseo (EMSIRVA) 1996, 1997, 1998; and Municipio de Santiago de Cali, 1999.

EMCALI and EMSIRVA have problems with collections, especially inside the municipal group. In 1998, the municipality did not pay EMCALI for the electricity used in public lighting. On the other hand, EMSIRVA, which has delegated the collection of its invoices to EMCALI, claims that the latter owes US$2.3 million for this item, and is preparing to directly collect its invoices.

In terms of personnel, the municipal companies, like the central administration, suffer from overstaffing. For each of EMCALI's services, for example, there are, on average, 102 customers per employee, when worldwide standards show that an efficient company should be able to cover 500 customers per employee. The tasks of customer services, invoice collection, human resources, accountancy, finance, management control, and general administration are carried out by independent teams in each of EMCALI's four service areas, when it would make sense to consolidate them in one team.

Due in part to the high number of employees and their favorable labor conditions, EMCALI and EMSIRVA have high operating costs. Both companies are applying policies to reduce personnel and rationalize operating expenses. Since 1998, EMCALI and EMSIRVA have begun to offer early retirement plans. EMCALI has decreased the number of employees from 4,100 in June 1998 to 3,100 in June 1999; of the 1,000 person reduction, 750 accepted early retirement plans and 250 retired and no replacements were rehired. EMSIRVA retired 300 persons in 1998 and is actively reducing the need for personnel through the use of private contractors. Both companies have an important pension burden as a result of the large number of retired persons: 3,200 in EMCALI and 909 in EMSIRVA.

In addition to personnel expenses, EMCALI has problems of water and electricity losses (36 and 16 percent of service consumption, respectively), partly due to technical reasons and partly to illegal connections. EMSIRVA, for its part, has been able to contain costs through outsourcing, recently transferring a new solid waste disposal dump to a private operator through concession. The company also foresees the substitution of its headquarters building with a less expensive space.

EMCALI and EMSIRVA operate with a small share capital, and their net worth is based on a recent asset revaluation. They are both highly indebted and need to refinance their debt to increase their share capital. In August 1999, EMSIRVA refinanced all its long-term debt. EMCALI has been studying the possibility of selling its electricity activity as a way to recapitalize itself and reduce its debt. By the end of 1999, EPM, the municipal company of Medellin, has been discussing with EMCALI the purchase of some capital shares for US$500 million.

The Metro project

In 1997, the public transport system of Cali consisted of 46 companies, operating 4,425 buses and 14,876 taxicabs. The proposed Metro project consists of a light rail system (LRS) crossing the central area of Cali, with 18.8 km of tracks and 19 stations, as well as the complementary infrastructure to support circular and radial bus lines. The project's own demand forecast is 178,000 passengers per day in year 2000 and 285,000 in 2010, or about 10 percent of municipality's 2.7 million population.[78] Given these numbers, the demand seems low and might translate into future operating deficits. By comparison, the Medellin LRS transports 235,738 passengers per day after four years of operation. In the first half of 1999, its operating revenues covered only 70 percent of the total operating cost.

Graph 7.7. Financial Adjustment Plan, 1998-2003:
debt including and excluding the METRO project (US$ million)

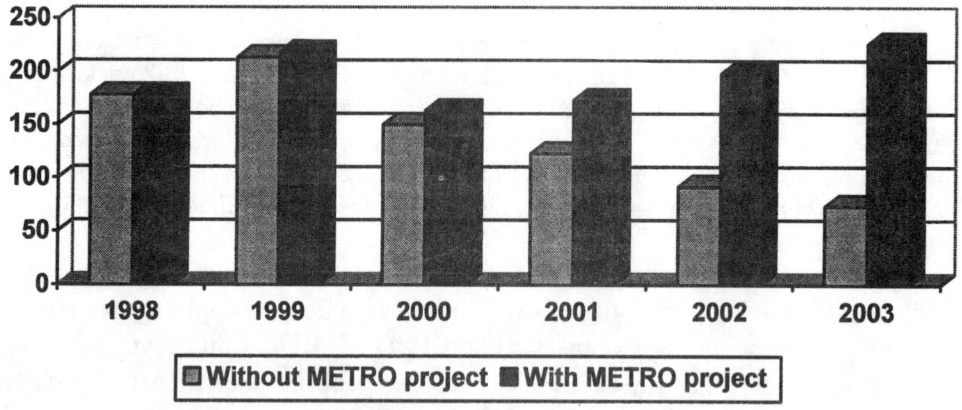

Source: Alcaldía de Santiago de Cali, 1999 (b).

The total investment for the Cali Metro is estimated at US$588.9 million, 31.5 percent to be financed by the municipality and 68.5 percent by the national government. In spite of a signed agreement specifying

---

[78] Urrea and Ortiz, 1999, estimate a decline to 2.45 million inhabitants in the year 2010, implying a smaller number of passengers.

this arrangement, however, there are problems with both financing sources. First, if the project is carried out, the Municipality of Cali would continue to have significant budget deficits in the next few years, unless the sale of EMCALI's business units takes place (please refer to Graph 7.7). According to the Financial Adjustment Plan, the municipal counterpart funds can be financed through: (a) a gas surcharge; and (b) new debt raised in the international markets. Given Cali's severe debt situation, it does not seem feasible for the municipality to access the international markets.

With regard to the national government, the public deficit was 3.3 percent of the GDP in 1998 and is estimated around 4 percent of the GDP in 1999. As a result, the 1999 budgeted investments amounted to US$3.7 billion, of which only US$1.0 billion had not been committed before 1999. Furthermore, the national government has been financing a significant portion of the Medellin Metro (the Municipality of Medellin has not been paying the debt in connection with its Metro and has invoked a guarantee given by the national government), and has also made a commitment to finance part of the Bogotá Metro. For these reasons, national government financing for the Cali Metro seems highly unlikely in the next five years.

Even though the project was launched in 1997 (to be completed by 2004), with the creation of an entity in charge of implementation, progress has been slow because of lack of funds. Supporters of the project claim that it will generate jobs needed in this time of economic recession. This might be true in the short run, particularly through construction activities, but in the long run, employment in the urban transport system will likely decrease as the LRS replaces part of the bus and taxi transportation system. <u>From all viewpoints, the Metro project comes out as an unfeasible option for Cali in the short run.</u>[79]

Proposed strategic priorities to solve the municipal finance crisis

The financial situation of the Cali municipal administration is difficult. The main problems can be summarized as follows: (a) expenditures are heavily concentrated in administrative and bureaucratic activities, rather than on social, infrastructure, and economic programs; (b) the central administration and the municipal companies are seriously overstaffed, making them highly inefficient, (c) tax rates, particularly for real estate, have increased importantly in the last years, without a concomitant improvement in public service delivery, thus generating frustration and distrust among citizens; (d) the forecast investment is centered around the Metro project, which is financially unfeasible and will, if undertaken, put off any substantial social investment for years to come; (e) the debt of the municipality is huge and its refinancing, as described in the Financial Adjustment Plan, seems unachievable. Given all of these issues, solutions to Cali's financial problems are related more to institutional modernization and human resource policies than to financing per se, and implementation of changes in these areas will be a difficult task. The thematic group's recommendations are:

• *Improve the municipality's credibility*. The Municipal Government has a low level of credibility among the citizens and the financial community. In this situation the first goal is to recover its credibility. In the case of the citizenry, the Municipality must commit to making its administration modern and transparent (this issue is covered at length in Chapter 2). In the case of the financial community, this means the commitment to implement a realistic financial plan for the next five years, showing results as early as possible. This implies the renegotiation of the Financial Adjustment Plan, because the commitment to prepay more than 50% of bank credits in the first three years is not realistic, except if an active policy of capitalization or sale of EMCALI is applied.

---

[79] The New Administration (2001-2003) completely stopped the Metro project

Since the Financial Secretariat in the Municipality does not have the technical skills needed to renegotiate the debt, and the Municipality lacks credibility, one option could be to create a restructuring board to deal with these issues. This board would be made of seasoned and well respected financial experts with high credibility and totally independent from political pressures. This might allow for more creative and aggressive solutions.

In addition, since the debt structure has a high interest risk exposure, the Financial Secretariat or the restructuring board has to increase its risk management capabilities, and ideally restructure the debt to fixed rate or to inflation indexed rate.

- *Increase operational efficiency.* Operational efficiency can be increased through three main activities: (i) improvement of human resource management; (ii) reduction of personnel; and (iii) contracting of certain services to the private sector.

The Human Resources Unit must be strengthened with adequate personnel able to apply consistent, transparent and centralized personnel policies, particularly with respect to definition of tasks, required professional profiles, re-assignment of personnel to different tasks, training to increase personnel capabilities, management by results, promotions, and negotiations with personnel representatives.

In terms of the quantity of staff, several measures are advised. First there should be a high degree of functional specialization that allows a reduction in the number of managers. Second, a greater process of decentralization of expenditures and of staff from the Central Administration to the districts, to strengthen the CALIs and the overall decentralized administration system, is advisable. Third there should be a policy of non-renewal of service and public contracts. Fourth, early-retirement plans should be offered and a policy of non-substitution of retired personnel must be applied.

Finally, transfer, through competitive bidding, of a significant part of the delivery of services to the private sector, is a desirable course of action. The municipality can promote and help to fund and manage associative companies with former municipal employees or small local companies that operate municipal services, generating employment in the city and reducing operating costs.

- *Increase revenues and investments through a small scale approach.* First, in terms of revenues, the policy has to focus on the improvement of tax collection, maintaining the level of tax rates. Bogotá is a good example of tax collection improvement through fiscal reform, fiscal information systems, and educational campaigns that can be applied in Cali.

Second, the sale of less productive assets, particularly land and buildings, is a positive policy but it must take into account the situation of the real estate and housing markets, in order to take the most advantage of it.

Third, investments must be small and strategic and hopefully will be based on the City Development Strategy. Small investments spread widely in all the city, focused on the improvement of the districts with the works made, basically, by local operators has to be considered.

- *The municipal companies.* From a strategic point of view, there are two consecutive courses of action for EMCALI: (i) to improve its efficiency; and (ii) to sell some of its business units.

Without the former, the latter might become unfeasible. EMCALI needs new investors to participate in its share capital as a way to capitalize the company, upgrade the services and reduce operating costs. Otherwise, private operators can compete in better terms than EMCALI, especially in the distribution of electricity and telecommunications. As a result, EMCALI can lose its market share, worsening the situation of the municipality. The capitalization or sale of EMCALI requires: (i) a clear regulatory framework in relation to the supply of services and price determination for low income customers; and (ii) a process of negotiation with the trade union representatives.

8. SUMMARY OF FINDINGS

Cali, like the rest of Colombia, faces the worst crisis of its recent history. Understanding this crisis has proven very difficult, as the problems in different sectors are intertwined and cannot be easily disentangled. This report is the result of an enormous effort to understand a city in crisis, and to formulate some initial recommendations for its recovery. This chapter synthesizes Cali's main problems and presents a series of propositions formulated jointly by the World Bank teams and the local counterpart teams.

Cali: a city in a severe crisis

First, Cali suffers from institutional devastation. Widespread corruption, brought on to a large extent by the boom in drug dealing activities, has caused a generalized decline in the trust that citizens and city stakeholders once had in their local leaders and institutions. Second, the city is in the midst of the worst economic crisis in its recorded history, with a GDP that has fallen more than 2 percent a year since 1995, and an unemployment rate of 20 percent. Although Colombia as a whole is in a severe economic recession, Cali is faring worse than other large cities in the country. Third, the social picture of Cali, a city that used to be a model of civic engagement and progress, is now bleak. Poverty has increased over the past years, with the latest figures showing a doubling between 1994 and 1998 of the percentage of the population living in misery. Inequality is rising, and the education system is not closing the gap between social classes or providing young people with the skills necessary to participate successfully in the labor market. In fact, the quality and coverage of education have worsened over the past 10 years. Violence, and particularly the rising number of homicides, has become one of the most important problems for Caleños and the most tangible sign of the decline in the city's livability. Fourth, the financial situation of the municipality is so critical, with a debt of US$196 million in 1999 and corresponding financial expenses at 200 percent of the gross current balance, that the very functioning of basic public services is at stake.

Proposals for the future

Despite the disheartening situation, the city still has many assets, particularly groups of concerned citizens and active entrepreneurs, some politicians of high integrity, a set of socially committed church leaders, a vigorous group of NGOs, and an academic sector that in spite of its own crisis is showing a desire for change and renewal. **The propositions presented below are the result of an intensive collaborative work between the World Bank and its local counterpart teams, and are by no means exhaustive, but represent rather an initial stimulus for change.** These propositions, originally presented in chapters 2 through 7, have been synthesized here and combined across sectors in some cases, in order to present a select and comprehensive menu. The propositions are separated into two groups: those that need to take place in the short run in order to spark some rapid changes; and those that need to be undertaken over the longer term because they require more resource investment and longer periods of time to produce tangible results. The short-term and long-term propositions are equally important.

Short term proposals

- *A pact among conscientious city stakeholders is the first condition for the city's recovery.* The situation of Cali is so dramatic in every sector, and the credibility of institutions is so eroded, that only a pact among stakeholders can help them to coordinate their positive actions. All

Caleños will have to make some sacrifices to begin the reconstruction process, and this requires a consensus around the following points: (i) recognition of the city's main problems, perhaps thorugh validation of this report and publication of a summary version by local newspapers, to stimulate discussion; (ii) selection of a limited number of specific short-term projects based on the proposals included in this report, and in the report produced by the NGO Foro Nacional por Colombia[80] in the context of the design of the Municipal Social Policy 1998-2000; (iii) agreement on measures to reform city government in terms of both staff reduction and composition, and on the design of incentives to create a culture of efficiency and responsiveness to clients; and (iv) development of a system of civil society sanctions against corruption. It is recommended that a neutral party act as a facilitator in the process, given the perverse climate of mutual distrust.

- *Improve the municipality's credibility, efficiency, equity, and revenues.* To improve the municipality's credibility among citizens, the municipality must commit to making its administration modern and transparent. The city might seek technical assistance in city management for this purpose. A serious review and evaluation of past governance initiatives is a first step in building a new administration. To gain the respect of the financial community, the municipality must commit to implementing a realistic financial plan for the next five years, and show results as early as possible. This implies renegotiation of the existing Financial Adjustment Plan, which is not feasible.

Since the municipality's Financial Secretariat does not have the technical skills, nor the city the credibility, needed to renegotiate the debt, one option could be to create a restructuring board made up of experienced and respected financial experts, and ensure that the board has independence from political pressures. This might allow for more creative and aggressive solutions. In addition, since the debt structure has high exposure to interest rate risk, the Financial Secretariat or the restructuring board has to have risk management capabilities, and ideally restructure the debt to a fixed rate or an inflation-indexed rate.

Operational efficiency can be increased through four main activities: (i) improvement of human resource management; (ii) reduction of personnel; (iii) contracting of certain services to the private sector; and (iv) reduction in the number of municipal programs, consolidation of the remaining programs under a limited number of strategic objectives, and improved coordination among institutions in their execution and evaluation. In terms of equity, the recommendation is to establish a progressive distribution of expenditures across the city across city sectors and services (right now the distribution is regressive), using poverty levels for adequate targeting.

Finally, to increase municipal revenues and investments, a small-scale approach is recommended. First, in terms of revenues, the policy has to focus on improved tax collection, while maintaining taxation rates. Second, the sale of less productive assets, particularly land and buildings, is a positive policy, but it should take into account the poor situation of the real estate and housing markets. Third, investments must be small and strategic and should ideally be based on the City Development Strategy, promoting small investments spread widely across the city, focusing on improvements in districts where works are controlled by local operators. At the same time, the metro project should be questioned, as it is impossible to finance in the current situation and is not the most important priority for Caleños. Fourth, the efficiency of EMCALI should be improved and some of its business units sold.

---

[80] FORO National por Colombia, 1998.

- *Generate employment through labor-intensive industries.* Two courses of action are recommended. The first one is to develop, with the Chamber of Commerce and relevant business associations, export strategies for Cali's most promising labor-intensive industries, namely shoes, garments, food processing, and paper and packaging. The second and complementary action is to expand efforts to develop training facilities, especially for young workers who have some education, and for migrants, particularly those located in squatter areas where facilities for training already exist. The municipality has a training program of this sort, but, because of budget limitations, it provides only partial coverage. The training should be related to identified sectors of labor-intensive manufacture and directed at exporting.

- *Develop a portfolio of select tradeable services.* The city has some clear strengths in terms of services. What needs to be done at this time is to improve a few of them. The recommendations are as follows: (i) in terms of finance, the planning authorities should be more proactive in encouraging the development of a financial quarter, as a key component of a central business district; (ii) for health, the Chamber of Commerce and national export agencies are already developing an export strategy (marketing, bank credit, transport, and accreditation) for the four areas of excellence in Cali's medical services, and the city needs now to identify clearly where these medical clusters will be located, their needs in terms of physical upgrading of facilities, and which ancillary services should be developed; (iii) in terms of higher education, Cali can use its considerable strengths to attract significant numbers of students, teachers, and researchers from outside the city, by and securing the collaboration of universities, colleges, and other institutions of tertiary education, and by providing complementary infrastructure such as theatres, concert halls, lecture halls, off-campus accommodations, stadiums, swimming pools, race tracks, etc.; (iv) in terms of culture, the city should support revitalization of the old town center by restoring its architectural and cultural heritage, guarenteeing security, making pedestrian malls, and landscaping its streets; and (v) in terms of information industries, the city needs to support the development of labor-intensive sectors (data loading, processing, etc.) by increasing the supply of specialized courses, including in squatter settlements.

- *Give special attention to youth.* Cali's youth are the most vulnerable group in the present situation. Several measures to increase attention to this group are proposed: (i) develop a comprehensive youth strategy for the city, with the active involvement of city stakeholders; (ii) seclude serious juvenile offenders in safe and appropriate correctional facilities; (iii) develop an outreach program directed at street gangs and other youth at risk; (iv) protect children from dropping out of school through family incentives, and the creation of better learning environments for children at home and in neighborhoods.

In addition, several programs can be undertaken to improve the city's youth labor supply. The first step is to thoroughly assess the city's labor supply to determine the major skills of the young productive force, and its potential for the future. Second, a program should be created to improve the contribution of education to the city's labor supply. This program would work toward better coordination of formal and non-formal education to develop the human resources needed by the productive sector, and facilitate an easy school-to-work transition for young people. The program would include: (i) reform of secondary schools, particularly in science, mathematics, and information technology; (ii) increased opportunities for secondary students to have hands-on experience in the use of advanced technologies (this program could be supported by private companies); and (iii) internships providing work experience in private and public organizations. Third, an identification of facilities and conditions needed for technological research would help the city to direct resources to promoting technological advances.

Long term proposals

- *Improve the city's infrastructure.* Cali needs the following infrastructure improvements: (i) completion of the Calima III hydroelectric project, and other projects, to ensure adequate energy supplies; (ii) completion of the rail link to the Buenaventura port; (iii) upgrading the road link to Buenaventura, particularly expanding the road from the point where freight movement from Cali coincides with that from Bogotá; and (iv) upgrading the airport and reducing air-freight charges. Ensuring adequate security of cargo handling at the airport provides a transport link that is less vulnerable to interruption than road or rail links. With such security, the following investments would make sense: a dry-dock (so that in-bond cargo can move either by air or via the sea ports), an intermodal junction point, and a business park and/or an export processing zone or free trade zone.

- *Improve the economic and social databases.* In the age of information, precise data on socioeconomic conditions are crucial for city management. At present, economic data at the city level are deficient, particularly for the informal economy, which seems to generate a significant number of jobs in the city, and for the tertiary sector, which constitutes two thirds of the city's economy. A unit dedicated to collecting and analyzing economic data for the city could be created with the help of academia and the Chamber of Commerce. Such a unit could also assess what transport infrastructure is needed to support economic development. At the time this documents was being revised for publication, several institutions in Cali joined together to create an Economic Observatory for the Valle del Cauca Department. The Economic Observatory has produced quarterly reports with an impressive amount of information on sector outputs, employment data, geographic location of employment and time trends. A special effort has been made to gather information on the informal economy.

However, the city still lacks reliable poverty data to monitor the impact of social programs. A monitoring unit, perhaps sponsored by private business and including representatives of civil society, should be established to evaluate the impact of city social policies, perhaps by using an instrument such as the EPSOC survey (see Chapter 4). Further, the monitoring unit could record, systematize, and publish institutional maps of the city (as is currently being done by the Universidad del Valle), which would allow for precise knowledge about the objectives, target groups, implementing agencies, and expenditures of existing programs. One of the first tasks of the monitoring unit would be to carry out an in-depth study of the problem of food insecurity, which was identified in the EPSOC survey, to pinpoint its extent and location. The same unit might also address economic and social aspects, thus constituting a think tank for the city.

- *Promote the inclusion of civic values in education.* Inclusion is a strategic priority. It helps to improve social cohesion, promote the participation of young people in the democratic process, and strengthen the personal values that support peace and democracy. Realization may involve teacher training on how to manage democratic schools and classrooms, and the formulation of learning standards related to solidarity skills, conflict resolution, and civic participation—all of which students should exhibit by the end of high school.

- *Create an education management mechanism.* Education is one of the largest and most critical enterprises in Cali. It requires a small, flexible, and learning-oriented management system, perhaps implemented by an experienced and respected NGO with the participation of relevant stakeholders. The main role of this NGO would be to propose municipal educational policies, provide reliable information on school demography and financial statistics, and assess school and teacher performance in order to guide public and private schools in improving the quality of their services. The main components of this management system could be:

- A continuous evaluation operation that would develop standards for students and schools and report to citizens and government. Such a system would replace current ineffective supervision and inspection units.

- A fund to make grants to research centers, universities, and private organizations for special studies and analyses.

- A monitoring and communication unit to provide information to individual schools, students, families, and citizens on the outcomes of schooling, and processing feedback to the system.

- Incentives to promote development of the city's learning industry (publishing and educational software companies, educational television, etc.), and to encourage the participation of churches, sports clubs, private industries, and mass media in education.

• *Improve the community's trust in local government.* To improve trust, the municipality should undertake small but visible programs in direct association with the direct participation of communities. One such program could be the recovery of public spaces, now severely deteriorated due to the invasion of delinquents, gangs, and drug dealers. The municipality could enter into a partnership with communities, civil society groups, local NGOs, and the police, helping to close the gaps among various local institutions. The program would include physical upgrading of public spaces and continuous and appropriate forms of law enforcement to prevent the return of crime and violence. Other options are support of the *Cali Centro* program with a similar tri-partite consortium, and the establishment (as mentioned in the first recommendation) of a civil society auditing mechanism to follow the performance of public institutions and impose sanctions for corruption.

ANNEX A. MAP OF CALI

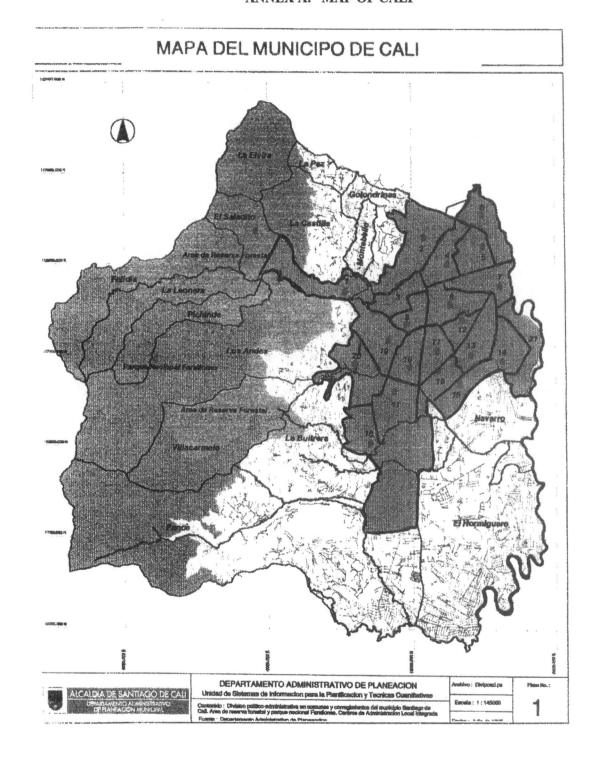

ANNEX B. CALI AT A GLANCE

1. Population
- Population in municipality in 1999: 2 million
- Average population growth in last 5 years: 1.83% per year
- Projected growth for next 10 years: 2.3% per year
- High rates of in-migration: in 1997, 45% of net population growth= in-migration
- Density: 156 inhab/ha

2 Economy
- Cali GDP = US$6 billion, 7% of national GDP (1996)
- Cali GDP grew by 9.21% in 1994 then decreased by 1.8% in 1995 and 4.7 % in 1996.
- By sectors: Manufacture: 24%, Communal services: 22.5%, Real estate services: 16%, Financial services: 11%, Construction: 11%, Commerce: 8.9, Agriculture: 0.4%

3. Employment
- Cali employs 14-16% of all the employed population in Colombia
- Employment by sector: Communal services: 30%, commerce: 24%, manufacturing: 20%, financial services: 7%, construction: 5%
- Unemployment rate: 10.8% in 1995, estimated 21% in 1999
- 52% of all employment was in the informal sector in 1994, 60% in 1998. Highest informality levels in commerce and construction

4. Poverty and inequality
- In 1998, 10% of population under misery level, 39% under poverty level.
- High inequality: Gini coefficient = 0.542, worst among large Colombian cities after Barranquilla (1998).
- Unemployment in lowest income quintile = 35.9% in 1999
- In lowest quintile, 83% of those employed work in the informal sector
- 15% of children between 12 and 16 years old were in work force in 1998

5. Violence
- 1980: 30 murders/100,000 inhab.
  1994: 124
  1999: 95
  (for comparison, in other large Latin American cities: San Salvador 84, Rio de Janeiro 61, Caracas 56, Sao Paulo 50, Mexico DF 9)
- 47% of all murders affect poor people

8. Municipal finance
- Main income items: real estate, industry and commerce taxes, and transfers from the National Budget
- Main expenses by program (excluding debt service): management and political direction, health and education, and transport and thoroughfares
- Main expense by type: personnel represents an 80%
- Total debt: US$196 million in 1999

ANNEX C. NOTE ON SPATIAL MANAGEMENT

Although spatial management was identified as the sixth issue to be addressed in the CDS, a full chapter on it was not prepared because: (i) the first comprehensive spatial plan for Cali—the *Plan de Ordenamiento Territorial,* POT, elaborated under the *Ley de Ordenamiento Territorial*—had just been issued elaborating at the time the World Bank mission visited Cali in September 1999, making an analysis premature; and (ii) the City Planning Department, because of its involvement in elaborating the *Ley de Ordenamento Territorial,* was not available to participate in the team meetings during that September, leaving the group without a key participant. The limited group that worked on this issue, however, did identify some problems and offer some proposals.

The main problems with the first comprehensive spatial plan for Cali are:

- Lack of a development model for the city that can clearly guide land use and infrastructure planning activities, including lack of an economic model and weak coordination with neighboring municipalities.
- Lack of a strategic framework, leaving the spatial plan to try to cover all the sectors that affect or are affected by space.
- Lack of a vision for the city that is shared by major elements of civil society

These problems notwithstanding, it is important to emphasize that the Cali Planning Department has made a strong effort in terms of data gathering and synthesis and cartography. The proposed plan needs now to be shared and discussed with civil society. The group recommends that the plan be maintained as a flexible tool and that the Planning Department consider the results of the CDS as an important input. Given this background, the strategic recommendations at this time are to:

- *Seek technical assistance in the diagnosis of selected issues, particularly housing, transport, and the environment.* Housing, transport, and the environment are three key aspects of the POT that have not been sufficiently studied. In the past, the city has had a series of housing programs that have had varying degrees of success, but there is neither a recent inventory of the housing stock nor a comprehensive evaluation of past programs. Transport is another key important element of the POT. The problem is that the city plans and the POT are both based on the assumption that a metro rail system, will be constructed. However, for financial reasons (please refer to Chapter 7), this project is unfeasible. Moreover, the recent Bank-financed household survey has shown that it is not a priority for Caleños (for details, refer to Chapter 4). Therefore, it is necessary to have alternative transport plans. Finally, in terms of the environment, there are several pressing problems, among which are the contamination of rivers and the close proximity of a large landfill. These issues, more than some others, need to be tackled in a regional manner, but the POTs of Cali and surrounding municipalities are rather independent from each other.

- *Promote regional coordination in terms of spatial planning with neighboring municipalities.* There have been some efforts by Cali to coordinate regional plans, but these efforts have been scattered and short-lived. Cali depends heavily on surrounding municipalities for industrial activity and employment. It is, therefore, in Cali's best interest to lead a regional coordination in terms of spatial planning.

- *Develop effective methods to consult with the community and encourage feedback regarding the implementation of the new spatial plan.* The City Planning Department is committed to this activity. Unfortunately, the consultation and feedback process in Colombia is subject to legal deadlines, and the information is not generally presented in an easily understandable form. Therefore, it is recommended that a serious effort be made to effectively disseminate the plan and discuss it publicly.

Bibliography

Alcaldía de Santiago de Cali. 1996. *Proyecto Epidemiología de la Violencia en Cali, 1993-96.*

———. 1998 (a). *Cali: Datos y Cifras.*

———. 1998 (b). *Cali, Municipio Saludable por la Paz, Un Proyecto Visionario.*

———. 1998 (c). *En Busca de la Equidad: Radiografia Social de Cali, Resumen.*

———. 1999 (a). *Cali Cívica y Pacífica. Programa Integral de Convivencia y Seguridad* (September).

———. 1999 (b). *Plan de Ajuste Financiero.*

Alcaldía de Santiago de Cali, Departamento Administratrivo de Planeación Municipal. 1998. *Plan de Desarrollo Económico y Social de Cali. 1998-2000.*

Alcaldía de Santiago de Cali, Empresa de Transporte Masivo de Cali. 1998. *Descripción del proyecto sistema integral de transporte masivo para el municipio de Santiago de Cali* (December).

Alcaldía de Santiago de Cali, Secretaría de Educación Municipal. 1998. *Plan Educativo Municipal: 1998-2000.*

Alcaldía de Santiago de Cali, Unidad Coordinadora Municipal (UCM), Programa de Apoyo a la Convivencia y Seguridad. 1999. *Informe de Ejecución* (first semester).

Alvarez, Benjamin. 1999. "City of Cali: The Education Perspective." Background paper for the World Bank's City Development Strategy. Washington D.C.

Alvarez, Benjamín, and Mónica Ruiz-Casares, eds., 1997. *Evaluación y Reforma Educativa: Opciones de Política.* Washington, D.C.: Academy for Educational Development/Inter-American Dialogue.

Atehortúa, Adolfo, Jose Bayona, and Alba Rodríguez. 1998. *Sueños de Inclusión. Las Violencias en Cali. Años 80.* Santa Fe de Bogotá: CINEP.

Atehortúa, Adolfo, Alvaro Guzman, Jaime Patiño, and Rodrigo Valencia de la Roche. 1995. *La Impunidad en Cali.* Cali: , Centro de Investigaciones Socioeconómicas de la Universidad del Valle.

Beaton, Albert E., Michael O. Martin, Ina V. S. Mullis, Eugenio J. Gonzalez, Teresa A. Smith, and Dana L. Kelly. 1996. *Science Achievement in the Middle School Years: IEA's Third International Mathematics and Science Study (TIMSS).* Chestnut Hill, MA: International Association for the Evaluation of Educational Achievement.

Becerra, Jaime. 1997. *Patrones de Salud y Pobreza en Areas Urbanas de Bajos Ingresos y Respuesta de los Servicios de Salud a la Luz de la Equidad.* Santiago de Cali.

Camacho, Alvaro. 1992. "Public and Private Dimensions of Urban Violence in Cali." in Charles Berquisst and others, eds., *Violence in Colombia.* Wilmington, Delaware: Scholarly Resources.

Cámara de Comercio de Cali. 1992. Programa Ciudadano "Cali que queremos," Resumen Ejecutivo (May), Cali, Colombia.

———. 1999. *Separata: Boletín de Noticias de la Cámara de Comercio de Cali* (various issues), Cali, Colombia.

Cámara de Comercio de Cali, Fundación para el Desarrollo Integral del Valle del Cauca. 1998. "Informe Monitor Situación Competitiva de la Región.". Cali, Colombia.

Castro, Manuel Fernando, and Manuel Salazar. 1997. "Acciones del Estado para Promover la Convivencia y la Seguridad en las Ciudades." in *Planeación y Desarrollo,* Vol. XXVIII, No. 4.

Comisión de Convivencia Ciudadana. 1997. "La Gestión de Paz." Cali Chamber of Commerce.

———.1999. "La Seguridad y la Convivencia en Cali: Plan de Contingencia a Corto Plazo." Cali Chamber of Commerce.

Comisión Vida, Justicia y Paz, Arquidiócesis de Cali. 1997. *Desplazados en Cali. Entre el Miedo y la Pobreza.*

Cubides, Fernando. 1998. "Cali (mas Yumbo, Jamundi, Candelaria, La Cumbre y Viajes)." en Fernando Cubides and others, *La Violencia y el Municipio Colombiano (1980-1997).* Santa Fe de Bogota: Universidad Nacional.

Dávila, C., A. Mejía, J. Paredes, and C. E. Bernal. 2000. *Ediciones Empresariales. Colciencias y Corporación Calidad.* Bogota: Tercer Mundo Editores.

Deas, Malcolm. 1999. "Violence Reduction in Colombia: Lessons from Government Policies over the Last Decade." mimeo. St. Anthony's College, Oxford.

Departamento Administrativo de Hacienda, Catastro y Tesorería, Municipio de Santiago de Cali. 1999 (a). "Deuda Pública por Entidades y Lineas de Crédito a 30 de Mayo de 1999 y a 10 de Junio de 1999." Cali, Colombia.

———.1999 (b). "Ejecución Presupuestal de Ingresos y Gastos a Diciembre 31 de 1998". Cali, Colombia.

———.1999 (c). "Informe Programa/Fuentes de Financiación" . Cali, Colombia.

———.1999 (d). "Presupuesto Cali 1999." Cali, Colombia.

Departamento Administrativo de Planeación. 1993. *Estudio Socioeconómico de la Ciudad de Santiago de Cali*. Cali, Colombia.

———. 1994. *Informe de Gestión, Junio 1992-Dic.1994*.Cali, Colombia.

———. 1995a. *Compartamiento del Sector Industrial de Santiago de Cali*. Cali, Colombia.

———. 1995b. *Plan de Desarrollo*. Cali, Colombia.

Departamento Administrativo de Planeación Municipal, Alcaldía de Santiago de Cali. 1999. *Cali en Cifras 1998*. Cali: Rodrigo Ordóñez Editores.

Departamento Administrativo Nacional de Estadística 1999.Statistics collected for this study, Bogotá, Colombia. www.dane.gov.co.

Departamento Nacional de Planeación. 1999. *Estadísticas*. www.dnp.gov.co.

Desepaz. 1998. *Atlas de Las Muertes Violentes en Cali – 1993-1997* (March).

———. 1999.Data collected for this study, Cali, Colombia.

Dillinger, William and Steven B. Webb. 1999. *Decentralization and Fiscal Management in Colombia*. World Bank Policy Research Working Paper 2122. Washington, D.C.

División de Administración y Ejecución Presupuestal, Municipio de Santiago de Cali. 1999. Data collected for this report, Cali, Colombia.

Echeverri, Oscar. 1998. "Cuántas Armas de Fuego hay en Cali? Bases para su Control." *Revista DESEPAZ* (Vol. 1, No. 1 (October).

El Tiempo. 1999. "Empeñado el Metro de Medellín." (September 12).

Empresas Municipales de Cali (EMCALI E.I.C.E.). 1996, 1997, 1998. *Estados Finacieros Anuales* de *EMCALI, ENERCALI, EMCATEL, ACUACALI y GENERCALI*.

———. 1999. *Ejecución Presupuestal Consolidada a julio de 1999 de EMCALI, ENERCALI, EMCATEL, ACUACALI y GENERCALI*.

Empresa de Servicio Público de Aseo de Cali (EMSIRVA E.S.P.). 1996, 1997, 1998. *Estados Financieros Anuales*.

———. 1999. *Estados Financieros a junio 30 de 1999*.

EPSOC (Encuesta de Acceso y Percepción de los Servicios Ofrecidos por el Municipio de Cali), background survey prepared for this study. Cali, Colombia, 1999.

Espitia, Victoria Eugenia. 1997. *Lesiones Fatales Intencionales y no Intencionales en Cali 1993-1996*. Santa Fe de Bogota, D.C.: Ministerio de Justicia y del Derecho.

———. 1998. "Epidemiología de la Violencia en Cali." *Revista DESEPAZ,* Vol. 1, No. 1 (October).

Filmer and Pritchett. 1999. "The Effect of Household Wealth on Educational Attainment: Evidence from 35 Countries.", *Population and Development Review (US)*, (March).

Foro Nacional por Colombia. 1998. *La Politica Social Municipal—Notas para la Definicion de un Enfoque*.

———. 1999. *En Busca de la Equidad—Radiografia Social de Cali*.

Franco, Saul. 1999. *El Quinto: No matar. Contextos Explicativos de la Violencia en Colombia.* Bogotá: TM Editores.

Fundación Carvajal. 1996. *Reporte Annual 1996.* Cali, Colombia.

Fundación para el Desarrollo Integral del Valle de Cauca. 1996. *Coyuntura Económica del Valle del Cauca* (January).

Gaviria, César. 1998. *Educación en las Americas: Calidad y Equidad en el Proceso de Globalización.* Washington, D.C.: Organization of American States.

Goldman Sachs. 1998. *The Emerging Markets Currency Analyst.*

Gómez, Hernando, ed. 1998. *Educación: La Agenda del Siglo XXI. Hacia un Desarrollo Humano.* Bogotá: Programa de Desarrollo de las Naciones Unidas, Tercer Mundo Editores.

Guzmán, Alvaro, coord. 1999. *Coyuntura Socio-económica Regional (Fase II). Informe Final.* Centro de Investigaciones Socioeconómicas de la Universidad del Valle, Universidad del Valle. Cali, Colombia.

Guzmán, Alvaro, and Marta Domínguez. 1996. *Diagnóstico de los Homicidios en Cali Durante 1996.* CIDSE, Universidad del Valle. Cali, Colombia.

Guzmán, Alvaro and otros. 1993. "Violencia Urbana y Seguridad Ciudadana en Cali." *Revista Foro* 22.

Harris, Nigel. 1999. "Towards an Economic Strategy for the City of Santiago de Cali." Background paper for the World Bank's Cali City Development Strategy. Washington D.C.

Hentschel Jesko, Kalpana Mehra, and Radha Seshagiri. 1999. "Poverty in Cali, Colombia." Background paper for the World Bank's Cali City Development Strategy. Washington D.C.

J. P. Morgan Securities, Inc. 1999. *Emerging Markets: Economic Indicators* (various editions).

———. 1999. *Emerging Markets Research* (September 29).

———. 1999. *World Financial Markets* (various editions).

Jerozolimski, Gabriel. 1999. "Poverty in Cali, Colombia: What do we know about it?" Background paper for World Bank Cali City Development Strategy. Washington D.C.

Klevens, Joanne. 1998. *Evaluations of Interventions to Prevent or Reduce Violence in Bogotá, Colombia,* (Junio).

Lanjouw, Peter, and Martin Ravallion. 1995. "Poverty and Household Size." *Economic Journal* 105.

Martin, Gerard. 2000. "Crime and Violence in Cali, Colombia- A Diagnosis and Policy Propositions." Background paper for the World Bank's Cali City Development Strategy. Washington D.C.

Ministerio de Hacienda y Crédito Público. 1999 (a). *Presupuesto General de la Nación, 1998-1999: Actualización* (May 3).

———. 1999 (b). *Situación Fiscal de los Municipios. Propuestas del Gobierno Nacional en la Búsqueda de Soluciones.* 1999. Santa Fe de Bogotá, D.C.: Encuentro de Alcaldes (April 13 and 14).

Ministerio de Hacienda y Crédito Público, Dirección General de Apoyo Fiscal. 1998 (a). *Lineamientos para el Análisis de la Capacidad de Endeudamiento de las Entidades Territoriales* (July).

———. 1999 (b). *El Estado de la Descentralización en Colombia* (May 27).

Misión Ciencia, Educación y Desarrollo. 1996. *Colombia: Al Filo de la Oportunidad – La Proclama. Por un País al Alcance de los Niños,* Tomo I. Bogotá, Colombia.

Misión Siglo XXI. 1999. *Análisis de la Economía de Cali con Enfasis en el Sector Industrial.* Background paper for the World Bank's City Development Strategy. Washington D.C.

Moser, Caroline, and Cathy McIlwaine. 1999. Background papers for World Bank's violence sector work in Colombia. Washington D.C.

Municipio de Santiago de Cali. 1997. Data collected for this report. Cali, Colombia.

Municipio de Santiago de Cali. 1999. *Plan de Desempeño* (June 10). Cali, Colombia.

Ortiz, Q. Carlos Humberto. 1998. "La Coyuntura Económica y de Empleo del Valle de Cauca."*Coyuntura Social* 9.

Oxford Analytica Daily Brief. 1999 (a). "COLOMBIA: Peso Predicament." (July 1).

Oxford Analytica Daily Brief. 1999 (b). "LATIN AMERICA: Risk Aversion." (August 26).

Pastrana, Andrés. 1999. *Plan Colombia. Plan for Peace, Prosperity, and the Strengthening of the State.* www.presidencia.gov.co.

Pinedo, Melba. 2000.Data collected for this report. Cali, Colombia.

Pinedo, Melba, and Maria Elena Suarez. 1996. *Las ONGs como Vehículos y Motores del Desarrollo en la Ciudad del Valle – Un Modelo de Desarrollo. Relación Gobierno-Sociedad Civil.* PROCALI (Julio).

Policía Metropolitana de Santiago de Cali. 1999. "Proyecto Escuelas Cívicas Juveniles y de Seguridad Ciudadana." paper presented at Desepaz (September). Cali, Colombia.

Policía Nacional. 2000. *Criminalidad 1999.* Bogotá, Colombia.

Presidencia de la República de Colombia, Departamento Nacional de Planeación. 1998. *Cambio para Construir la Paz. Plan Nacional de Desarrollo 1998-2002.*

Roa, Carmen. 1997. "Mujer y Pobreza: entre la Pobreza y la Miseria." Fundación Foro Nacional por Colombia, Capitulo Regional del Valle de Cauca, Cali.

Rojas, Fernando, Alexandra Ortiz, and Melba Pinedo. 2000. "Institutional Reform in Cali." Background paper for the World Bank's Cali City Development Strategy. Washington D.C.

Rothschild, N.M. and Sons. 1999. "Preliminary Information Memorandum: Concession of the Alfonso Bonilla Aragon Airport of the City of Palmira." Cali: Advisory Group.

Rubio, Mauricio. 1996. *Crimen sin Sumario, Análisis Económico de la Justicia Penal Colombiana.* Bogotá: CEDE, Universidad de Los Andes.

———. 1999. *Crimen e Impunidad.* Bogotá: Tercer Mundo Editores.

Santamaria, Mauricio. 1999. "Poverty in Cali – Basic Comparisons and Developments." Background paper for the World Bank's Cali City Development Strategy. Washington D.C.

Secretaría de Fomento Económico y Competitividad. 1998 (a). *Cuentas Económicas de Santiago de Cali Métodos y Resultados.*

———. 1998 (b). "Industria Manufacturera de Cali" (October).

———. 2000. *Informe del Estudio de Reactivación Económica de los Ingresos y el Empleo para la Ciudad de Cali – Primera Fase.*

Servicio Nacional de Pruebas. 1999.Data collected for this report, Bogotá, Colombia.

Solans, Pilar. 1999 "The Municipality of Cali: Financial Review." Background paper for the World Bank's Cali City Development Strategy. Washington D.C.

UNESCO, 1999.Data collected for this report, New York, USA.

Unidad de Recursos Humanos, Municipio de Santiago de Cali. 1999. Data collected for this report. Cali, Colombia.

United Nations Development Programme. 1994. *Human Development Report 1994.* New York: Oxford University Press.

Universidad del Valle de Cauca. 1996. *Migration in the Municipality of Cali.* Cali: Institute of Pacific Studies.

Urrea Giraldo, Fernando. 1997. "Dinámica Socio-demográfica, Mercado Laboral y Pobreza Urbana en Cali Durante las Décadas de los Años 80 y 90." *Coyuntura Social* 17. Cali: Fedesarrollo and Instituto Ser de Investigación.

Urrea Giraldo, Fernando, and Carlos Mejia. No date. "Culturas Empresariales e Innovación en el Valle del Cauca." mimeo. Centro de Investigaciones Socioeconómicas de la Universidad del Valle. Cali, Colombia.

Urrea Giraldo, Fernando, and Carlos Mejia. 2000. "Innovación y Cultura de Las Organizaciones en el Valle del Cauca." en *Innovación y Cultura de Las Organizaciones en Tres Regiones de Colombia,* F. Urrea Giraldo and C. L.G. Arango, eds. Centro de Investigaciones Socioeconómicas de la Universidad del Valle. Cali, Colombia.

Urrea Giraldo, Fernando, and Carlos Ortiz. 1999. "Patrones Sociodemográficos, Pobreza y Mercado Laboral en Cali." Background paper for the World Bank's Cali City Development Strategy, Washington D.C.

Urrutia Montoya, Miguel. 1998. *Causas de la Recesión y Estrategia de Reacción*. Bogotá: Banco de la República.

Vanegas Muñoz, Gildardo. 1998. *Cali Tras el Rostro Oculto de las Violencias*. Universidad del Valle. Cali, Colombia.

Vargas Hernán, Bernardo Toro, and Martha Rodríguez. 1992. *Acerca de la Naturaleza y Evolución de Los Organismos no Gubernamentales en Colombia*. Bogotá: Fundación Social.

Velasquez, Fabio. 1996. "Una Mirada desde Cali." en *Nuevas Formas de Participación Política*. Bogotá: FESCOL.

Velez Ramirez, Humberto et al. 1998. "El Desarme: Por qué Decirle "si" en Teoría, Perspectiva y Realidad?" *Revista DESEPAZ*, Vol. 1, No. 1 (October).

Villaveces, Andrés. 1998. "Efecto del Plan Desarme en la Tasa de Homicidios en Santiago de Cali." *Revista DESEPAZ*, Vol. 1, No. 1 (October).

World Bank. 1997. *Country Assistance Strategy of the World Bank Group for the Republic of Colombia*, Report No. 17107-CO. Washington D.C.

———. 1998. *Colombia: Economic and Social Development Issues for the Short and Medium Term*, Report No. 18394-CO (November). Washington D.C.

———. 2000. *Violence in Colombia – Building Sustainable Peace and Social Capital*. World Bank Country Study. Washington D.C.